Applicative Chemistry of Tanning Metallic Heterocomplexes

Authored by

Carmen Gaidau

Leather and Footwear Research Institute
Bucharest
Romania

CONTENTS

Foreword   *i*

Preface   *iii*

CHAPTERS

1. **Understanding of Leather Tanning**   3

2. **Chromium in the Leather Industry and its Environmental
    Implications**   20

3. **Synthesis and Use of Tanning Metallic Heterocomplexes**   36

4. **Theory and Experimentation of Synthesis Reactions of Tanning
    Metallic Heterocomplexes**   50

5. **Structural Analysis of Tanning Metallic Heterocomplexes and
    Testing their Tanning Properties**   95

    References   123

    Index   130

FOREWORD

Since prehistoric times, humans have used animal skins for clothing, to wear and to elaborate products with different uses.

The tanning industry is characterized by exploitation of natural resources and a constant interaction with the environment. In fact, animal skins are a byproduct of the food industry. If these skins wouldn't be processed, they become a waste to be treated and eliminated causing health and environmental problems.

Throughout history different methods of tanning that are no longer used today had been applied, that is the case of tanning using smoke. However, other methods are still used today, such as aluminum salts, a system widely employed in ancient Rome. Also, the tanning by fats was known in the pre-classical Greece and is currently used to obtain chamois. But the use of these systems represents only 1% of the current production of skin.

The only traditional system (used by Egyptians, Romans and Greeks) that maintains a high importance is the vegetable tanning. However, just represents the 10% of world production, with a tendency to regain importance.

In fact, the chrome tanning method is the most widely used and represents the 90% of world production, thanks to its ease of use and good properties that gives to the skin: durability, hydrothermal resistance, touch, fullness, *etc*. However, the use of chromium is a controversial issue because of its toxicity and persistence in the environment that represent some of its chemical forms.

A great variety of projects have been carried out in order to minimize this impact: recycling of pickle-tanning floats, recovery and treatment of chrome floats, high exhaustion of such floats, management of solid waste containing chrome, and the use of other tanning agents to substitute chrome.

Indeed, the volume *Applicative Chemistry of Tanning Metallic Heterocomplexes* includes essential background about the leather tanning process and the use of chromium and its environmental implications. In addition, the eBook reports the synthesis and use of tanning metallic heterocomplexes with the aim to provide the classic features of mineral leathers and at the same time to reduce environmental pollution.

This eBook is a good tool for researchers and experts of the leather industry to understand the multiple interconnection ability of functional groups of the collagen macromolecule and the action synergism of several types of tanning metals. In this eBook you will find a low-cost alternative to the chrome tanned leather by preserving their well-known characteristics.

Anna Bacardit Dalmases
Igualada Engineering School (UPC)
Catalonia
Spain

PREFACE

The volume ***Applicative Chemistry of Tanning Metallic Heterocomplexes*** depicts a comprehensive picture of a new chapter in the chemistry of coordination compounds, with industrial applications in the field with the longest history, natural leather processing. The chemistry of heterocomplex compounds is a fascinating field for experts in chemical synthesis and structural analysis, and for technologists specializing in leather processing. The volume describes the vast theoretical and practical possibilities of exploiting the action synergism of metals with different collagen crosslinking capacity. The possibility of reducing chromium content from leather tanning agents by replacing it with other tanning metals has significant environmental implications and minimum changes in terms of quality and production costs of natural leather, and is a viable alternative for a safe future.

The volume is intended for researchers, chemical auxiliary producers, experts in natural leather processing who are looking for clean and efficient solutions for wastewater pollutants, sludge or solid wastes while striving to preserve the known characteristics of mineral tanned natural leather.

Chapter 1 provides an overview of the history of natural leather processing, which is intermingled with the eras of human civilization development and helps to understand the chemistry of processing a complex material, unmatched by any other similar synthetic material, in terms of hygienic properties and compatibility with the human body. A venture in the modern era of basic chromium salts tanning helps to understand the relationship between types of chromium compounds complexed with various organic ligands, as well as the properties of leather. The possibilities of modeling sensory and physical-mechanical characteristics of leathers by using tanning agents with various properties suggests an infinite area of creativity which carries until modern times, when the search for ways to optimize organic tanning variants is still in progress.

Chapter 2 discusses the experience accumulated in the modern period regarding minimization of environmental impact of using basic chromium salts as tanning agent within the restrictions imposed by the aqueous reaction medium and the need to preserve natural properties of collagen fibers. Although scientific progresses in chromium salt management used in natural leather tanning and the long-term future of this material are acknowledged, the large chemical auxiliary companies are considering organic alternatives of collagen crosslinking. Resorting again to methods of optimizing tanning processes using metal salts is an increasingly discussed variant for restoring the known mineral character of traditional leathers. In this context, tanning metallic heterocomplexes are an intelligent, low-cost and much more versatile alternative compared to any other proposed variant.

Chapter 3 summarizes the attempts of heterocomplexing various tanning metals, and of understanding the structure and correlation with the stability and properties imparted to leathers. The multiple interconnection ability of functional groups of the collagen macromolecule and the action synergism of several types of tanning metals represent the argument that supports and promotes tanning metallic heterocomplexes as an alternative for the future.

Chapter 4 is a culmination of practical pilot scale experimentation activity, under industrial conditions, conducted by two authors who have decoded the complexing mechanism of three tanning metals with the potential of substantially reducing the use of chromium as tanning agent. Defining the stability boundaries of binary and ternary compositions of the new metallic heterocomplexes and the new calculation formulas to establish the synthesis reagents constitute a progress in the coordination chemistry of tanning agents. Chapter 4 can serve as a reference eBook for the synthesis of stable tanning metallic heterocomplexes within the technological limits of application in leather tanning, an original contribution in the field and a starting point for any new research endeavor.

Chapter 5 deals with specific analyses for coordination complexes, for the purpose of anticipating the behaviour of tanning metallic heterocomplexes upon the interaction with natural leather. Analysis of ionic components explains the more pregnant anionic nature compared to that of the best known tanning agent, basic chromium salt, and therefore, the improved efficiency of using tanning metallic heterocomplexes in natural leather tanning.

ACKNOWLEDGEMENTS

The research was partially funded by Executive Agency for Higher Education, Research, Development and Innovation Funding (UEFISCDI) from Romania, under the project 167. The authors are grateful to Ms. Dana Gurau for the English translation of the manuscript.

CONFLICT OF INTEREST

The authors report no conflicts of interest in this eBook.

Carmen Gaidau

Leather and Footwear Research Institute Bucharest
Romania
E-mail: carmen_gaidau@hotmail.com

Applicative Chemistry of Tanning Metallic Heterocomplexes

2

Send Orders for Reprints on reprints@benthamscience.net
Applicative Chemistry of Tanning Metallic Heterocomplexes, 2013, 3-19

CHAPTER 1

Understanding of Leather Tanning

Abstract: Leather tanning is one of the oldest human activities, which intelligently exploits an animal by-product, natural leather. Although lost in the mists of history, the evolution of leather processing is closely related to the history of humankind and scientific progress. Types of tanning practiced by mankind are closely related to intuitive knowledge about conservation of organic, putrescible materials and chemical materials which interact with the former. Natural leather processing is presented in conjunction with the evolution of human society and historical epochs. Understanding the leather tanning process is closely related to understanding the interactions between different functional groups of collagen and crosslinkers: vegetable tanning agents, aldehydes or chromium salts. The connection between the value of using leather, its properties and the method of processing is the foundation of understanding natural leather processing technologies and the start of a journey into the fascinating world of alternative ecologic processing and chemistry of mineral tanning agents.

Keywords: Leather processing, Leather history, Collagen crosslinking, Mineral tanning, Vegetable tanning, Chromium basic salts.

BRIEF HISTORY

Leather processing emerged as a natural consequence of hunting and food, which is still valid nowadays. The first tools made of stone, bone or iron were probably intended for cutting and fleshing game [1].

Primitive humans had quickly learned that covering themselves with animal fur protects them from cold and helps them dominate their adversaries, beasts of prey. From this point on, man's preservation instinct and intuition have led him towards innovation and evolution.

Cavemen must have quickly learned that the leathers they wore could be fumed or treated with various natural substances in order to preserve them and maintain their softness and flexibility. Smoke and animal fat were the most readily available materials with preservation properties used by the primitive human to process the hides of hunted animals.

Drawings in a cave near Tervel, Spain depict women dressed in fur jackets and skirts; other cave drawings in the Spanish Pyrenees show men wearing fur boots and trousers. It is estimated that these drawings date back over 20,000 years [1-5]. Egyptians processed hides so well that there is evidence over 3000 years old still in perfect condition. Mummification art blends with that of leather processing. Egyptians and Jews have been recording many methods of tanning for the past 5000 years. Sumerians, Babylonians, Assyrians and Arab peoples were well acquainted with hide tanning using alum and plants rich in tannins. A Babylonian document describes a technique of tanning: "Take the skin of an ox and put it to soak in water. Add flour, beer and wine. Then put the skin in a pot with oil, essences from plants and process the skin using wheat flour. Then set the skin on alum stones from Hittites" [1].

The relatively recent discovery of the Ötzi iceman [6] has enabled identification using modern methods [7] of the type of tanning practiced in the Copper Age (3300 BC) by the population in the Alpine region. Goat, bear and deer fur skins tanned with fat used to be Ötzi iceman's basic clothing [5]. Ötzi iceman's footwear is considered to be the oldest footwear discovered [6], made of bearskin soles, deer skin upper, leather strings and having a functional, good quality design. Numerous clothing accessories of the Ötzi iceman are made of leather, which demonstrates the importance of this material in everyday life of the prehistoric man (Fig. **1**).

Figure 1: Ötzi the Iceman, 3300 BC [www.iceman.it].

The ancient Greeks and Romans have contributed greatly to the science of tanning, and some of their methods are still used. The word "tan" whose meaning takes its origin from oak bark (*tannum*) was invented by the Romans who used leather as currency.

Tanning was a popular activity in the ancient world, and methods for obtaining quality leathers were preserved and transmitted from generation to generation. By the mid-19th century, nearly all leathers were tanned with vegetable tannins.

Heavy hides were placed into basins and layers of bark, leaves and fruit of various plants were placed on top. Water was added gradually to fill the basin. Water extracted tannins from the vegetable materials that tanned skin. This process took months or even years to complete, especially if heavy cow or horse hides were processed. Between the ancient and renaissance period, vegetable tanning processes remained almost unchanged. In a 1564 proclamation, the Medici royal court established a period of six months as the minimum time of immersion of hides in tanning vessels, on the premises of the kingdom, where four centuries later, the largest tanning center in the world would be developed, on the banks of the river Arno [5].

Potassium alum was the next tanning agent that the ancients had discovered, especially people in the Middle East. Assyrians, Babylonians and Sumerians were known to be skilled in its use; the Greeks used this tanning method before the year 450 AD. Alum tanned skin is white and absorbs colours well; Egyptians dyed the skin in beautiful tones of red, yellow, blue and black using dyes from plants. In Spain, in the 18th century, the Spanish or Cordoba leather was made, which was either pure white or red and was famous throughout the world. Only in the 19th century, in Britain, was the introduction of leather tanned with alum reported as an important innovation. The main drawback of alum tanned leather is that alum is soluble in water and, if the skin is soaked in water long enough or repeatedly, alum is removed and the skin stiffens and begins to decompose.

No important innovations in leather processing have been reported until the 1880s, when a corset manufacturer from the United States complained to his leather supplier, August Schultz, that alum tanned leather causes corrosion of metal supports in women's corsets. He was complaining to Schultz because the latter was an experienced textile dyer who was familiar with dichromate-based mordants used to fix dyes. Schultz and his friends began to investigate a tanning process that was similar to alum tanning, but using chromium alum instead of potassium alum as a tanning agent. They finally found that the chromium solution could be fixed with soda ash and salt and could tan as well as potassium alum, with the advantage that it was not extracted by washing in water or sweating [2].

This very good tanning method is nowadays used for most skins, as a quick tanning version. Present-day tanneries are extremely efficient and, although they follow the same steps as traditional tanneries, some mechanical operations such as fleshing, splitting, shaving and drumming are carried out by machines. A qualified chemist may use various specialized substances which he must combine in right proportions, depending on the raw material and the characteristics of the finished product.

LEATHER TANNING, THE ART OF PROCESSING A BIOPOLYMER – THE COLLAGEN

Definition

Tanning is a complex process by means of which animal hide, specially prepared, is treated with chemical materials capable of crosslinking collagen, of reducing water amount in the skin, from approximate 80% in raw state, to 16% in finished leather state. The purpose of these complex treatments is to give this protein-based material the following characteristics:

- Stabilization upon the action of enzymes generating degradation bacteria and chemical materials;

- Increasing hydrothermal resistance;

- Reducing or eliminating the swelling tendency;

- Increasing mechanical resistance;

- Reducing density through separation of fibres;

- Reducing deformability;

- Reducing the contraction in volume, surface and thickness and

- Increasing porosity of fibrous texture [8].

The most important influence in achieving these properties is that of crosslinking agents of collagen represented by various tanning agents.

The Mechanism of Collagen Crosslinking Using Various Tanning Agents

Vegetable tanning materials [9] crosslink the collagen molecule through numerous hydrogen bonds forming between phenolic groups of the vegetable tannin and the amino groups of the collagen (Fig. **2**).

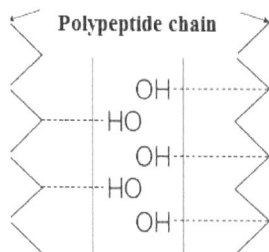

Figure 2: Hydrogen bonds between phenolic groups of the vegetable tannin and the amino groups of the collagen.

Mineral chromium tanning agents form covalent bonds between basic chromium groups and carboxyl groups of collagen (Fig. **3**).

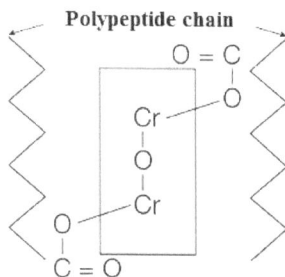

Figure 3: Coordination of carboxyl groups of collagen to the basic chromium complex.

Aldehydes, such as glutaraldehyde, crosslink collagen through covalences forming between the carbonyl groups of aldehydes and the amino groups of collagen (Fig. **4**).

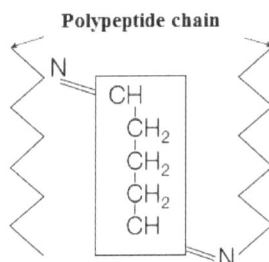

Figure 4: Covalent bonds between aldehyde groups and amino groups of collagen.

Synthetic tannins form electrovalent bonds between sulfonic acid groups of the synthetic tannin and the amino groups of collagen (Fig. **5**).

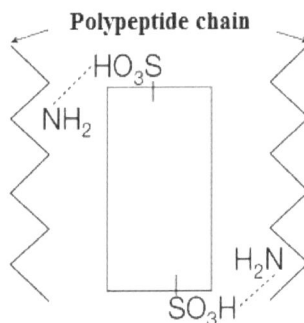

Figure 5: Secondary electrovalent bonds between sulfonic groups of the synthetic tannin and the amino groups of collagen.

Shrinkage temperature of collagen is the main characteristic indicating the crosslinking degree and, therefore, the stability of leathers to various types of aggression: mechanical, biological, chemical and physical.

Determining shrinkage temperature provides the first clues regarding the type of tanning. Thus, identification of the old types of tanning was done by comparing shrinkage temperatures of leathers tanned using known agents with those of historical leathers [7].

Shrinkage temperature value (Table **1**) for raw hides and processed skins is presented below [9].

Table 1: Shrinkage temperature for hides and leathers.

Material	Shrinkage temperature	Maximum temperature
Mammal collagen fibres	62 – 64 °C	37 – 38 °C
Fish collagen fibres	40 – 45 °C	25 – 30 °C
Pelt	40 – 60 °C	37 – 38 °C
Chamois leather	65 – 70 °C	40 °C
Aluminium tanned leather	70 – 75 °C	45 °C
Vegetable tanned leather	70 – 85 °C	45 °C
Formaldehyde tanned leather	80 – 85 °C	50 °C
Glutaraldehyde tanned leather	75 – 85 °C	50 °C
Aldehyde/aluminium tanned leather	80 – 90 °C	55 °C
Chromium tanned leather	100 °C	60 – 80 °C

Confirmation of tanning agents required complementary, non-destructive investigations using advanced FM-IR (Frustrated Multiple Internal Reflectance) techniques, X-ray fluorescence *etc*. In order to identify the type of tanning for Ötzi iceman clothing, who lived in the Copper Age (3300 BC) the results presented below were found (Table **2**).

Table 2: Identification of fats as tanning material along history.

Age	Type	Fat
11th century	Footwear uppers, vegetable	traces
12th century	Footwear uppers, vegetable	few
14th century	Sole leather, vegetable	very few
New, glove leather	Glacé leather tanned with aluminium	mixed fats
New, chamois	Deer skin, German	esters of fatty acids
New, reindeer	Reindeer skin, processed by Native Americans, fumed	very few
New, fish	Fish skin, Siberian	mixed fats
Millennium 4 BC (Copper Age)	Bearskin, deer skin	unidentified fat derivatives

Tanning methods can be categorized as follows:

Vegetable tanning:

- Basin tanning

- Accelerated tanning

- Quick tanning

Mineral tanning:

- Tanning with chromium salts (wet blue)

- Tanning with aluminium

- Tanning with zirconium

Other types of tanning, combined tanning:

- Tanning with aldehydes (wet white).

- Tanning with aldehydes and oils, chamois.

A classic technology flow is presented below (Fig. **6**).

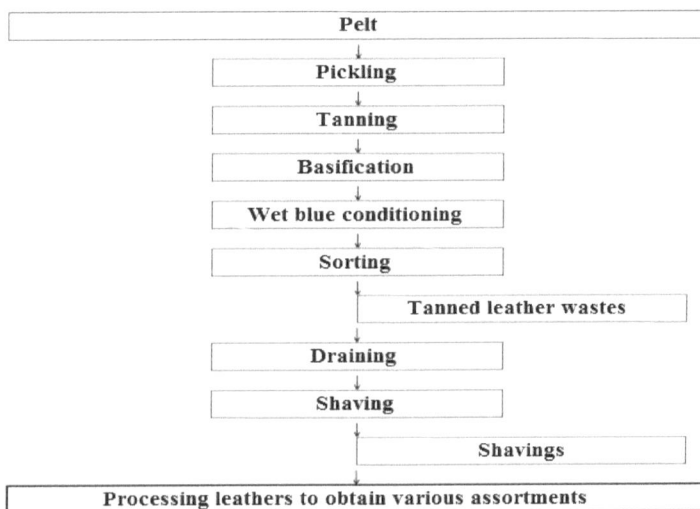

Figure 6: Classic technology flow.

The type of tanning decides the final appearance of leather. Vegetable tanning is one of the oldest types of tanning and is currently used for certain types of leather, such as leathers for soles.

Currently, about 80% of leathers are produced worldwide using chromium as tanning agent. An increasing amount of leathers, especially for car upholstery, is tanned with a combination of aldehydes and synthetic tannins, vegetable tannins or polymers.

Leather retanning is of particular importance in leather processing; it allows increasing the value of leather. Syntans, resins and polymeric agents are tanning agents applied in the retanning process in order to correct leather characteristics and to improve their quality.

The classic mineral tanning process of retanning hides after neutralization, which raises the pH of skins after tanning to allow an even distribution of retanning materials, is not applied in the case of purely organic tanning. Basically, a strict distinction can no longer be made between tanning and retanning.

Quantities of tanning agents used in order to make various types of leathers are presented below.

Quantities of tanning agents needed for different types of leathers (percentages in relation to leather weight):

Vegetable/Synthetic Tanned Leathers

Sole leathers: 33 – 40% pure tannin

Inner sole leathers: 25 – 30% pure tannin

Combined tanned leather: 30 – 33% pure tannin

Heavy leathers, technical leathers: 28 – 30% pure tannin

Handbags and upholstery leather: 20 – 25% pure tannin

Upper leathers: 20 – 25% pure tannin

Sheep and goat leathers: 15 – 20% pure tannin

Light leathers and linings: 2-12% pure tannin

Splits: 12 – 18% pure tannin

Mineral Tanned Leathers

Chromium tanned leathers: 1 – 4% Cr_2O_3

Aluminium tanned leathers: 1 – 8% Al_2O_3

Aldehyde tanned leathers: 2 – 8% aldehydes

Chamois leathers: 25 – 40% fish oil.

Other Tanning Methods

Variant A: without heavy metals and aluminium salts

Variant B: without chromium, but using Al, Zr, Ti, Fe.

The wet white process, carried out by means of purely organic tanning can be done using aldehydes, vegetable or synthetic syntans, polymers, various auxiliaries. Leather properties are adjusted particularly by the amount and type of tanning agent and by process parameters. Thus, a great variety of leathers can be achieved: automotive leather, upholstery leather, leather for clothing and for footwear uppers. Wastes resulting from shaving and trimming leathers are chromium-free.

The technology flow diagram for wet white leather production is presented below (Fig. **7**).

The most important parameters in making wet white leather [10] are:

-   complete deliming and bating of pelt;

- intense washing;

- pickling across the skin section;

- type and concentration of aldehydes;

- pH;

- stirring time;

- basification and

- using syntans.

```
┌─────────────────────────────────────────────────────────┐
│                          Pelt                           │
└─────────────────────────────────────────────────────────┘
                            ↓
            ┌─────────────────────────────────┐
            │        Deliming/Bating          │
            └─────────────────────────────────┘
                            ↓
            ┌─────────────────────────────────┐
            │        Pickling/Tanning         │
            └─────────────────────────────────┘
                            ↓
            ┌─────────────────────────────────┐
            │        Draining/Shaving         │
            └─────────────────────────────────┘
                            ↓
                  ┌──────────────────────────────┐
                  │           Shavings           │
                  └──────────────────────────────┘
                            ↓
┌─────────────────────────────────────────────────────────┐
│                     Tanning with:                       │
│  Vegetable or synthetic   │                             │
│  tanning agents or polymers │   Basic chromium salt     │
└─────────────────────────────────────────────────────────┘
            ↓                            ↓
┌─────────────────────────────────────────────────────────┐
│        Processing specific to various assortments       │
└─────────────────────────────────────────────────────────┘
```

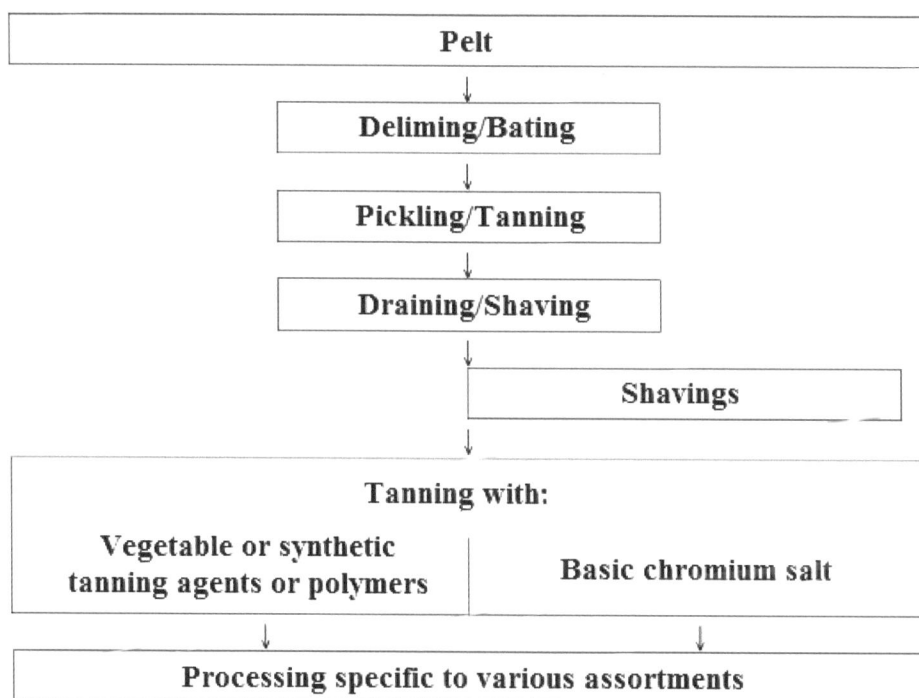

Figure 7: Technology flow diagram for wet white leather production.

Oil Tanning

This type of tanning is special, using non-saturated oils and producing very soft leathers, named chamois.

Tanning/Retanning

Tanning is an operation by means of which a biopolymer, the collagen of hides, is stabilized and it can be done by using vegetable, mineral or synthetic tanning materials. Vegetable or synthetic tanning materials are generally anionic materials and act similarly on the hide protein. Chemically, they are complex mixtures, and their means of action is still not well known. Vegetable tannins can be categorized into two main groups:

1) Hydrolysable tannins – complex esters of phenol and carboxyl and polyhydroxy phenolic acids;

2) Condensed tannins – mixtures of products with various degrees of polymerisation, containing various catechin molecule or catechinic compounds.

Syntans are preponderently condensation products of sulfonic acids (phenolic or naphtenic) with formaldehyde or similar structures.

Vegetable and synthetic tanning agents are frequently used to modify the properties of leathers tanned with chromium salts or to produce leathers for soles, belts, balls or other types of leathers of which elasticity or softness is not required.

USING CHROMIUM IN LEATHER TANNING

The most important mineral tanning agent is trivalent chromium sulphate, widely used to produce leathers for clothing and footwear uppers, in fact, all leathers of which softness and elasticity are required. The synthesis reaction of chromium tanning salts is the following:

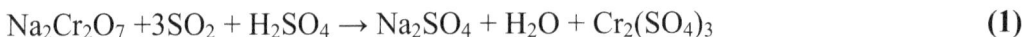

$$Na_2Cr_2O_7 + 3SO_2 + H_2SO_4 \rightarrow Na_2SO_4 + H_2O + Cr_2(SO_4)_3 \tag{1}$$

Chromium is a transitional element which forms coordinative complexes by using 3d orbitals to attract external electrons and it is not matched so far by any other element. The complex is basic because it contains hydroxyl groups coordinated to the chromium atom. Two of the main characteristics of the chemistry of chromium impart tanning properties to the latter. The first characteristic consists in average stability of formed complexes, which have the ability to change ligands relatively easy. The second characteristic consists in the ability of chromium to form polynuclear complexes in which Cr-O-Cr bridges are involved.

Basic Chromium Salts, oLo and oxo Compounds

The final process which basic chromium salts undergo is oxolation, which is a slow and almost non-reversible process. Thus, long chains of chromium complexes are formed. These polynuclear compounds, which penetrate the gaps between macromolecular chains of skin collagen and can crosslink by constructing bridges, are the essence of the tanning phenomenon.

These two basic properties of chromium can be found individually in other elements, which do not have particular tanning properties. The two properties together confer the basic tanning character. Chromium shows a strong tendency to form coordination compounds with molecules containing carboxyl groups:

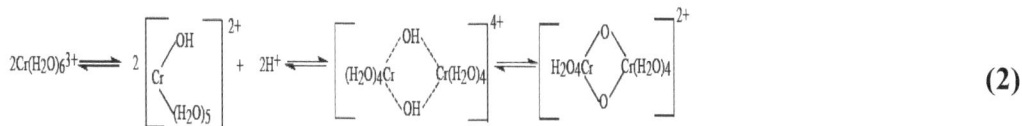

$$2Cr(H_2O)_6^{3+} \rightleftharpoons 2 \begin{bmatrix} Cr \underset{(H_2O)_5}{\overset{OH}{\diagup}} \end{bmatrix}^{2+} + 2H^+ \rightleftharpoons \begin{bmatrix} (H_2O)_4Cr \underset{OH}{\overset{OH}{\diagdown\diagup}} Cr(H_2O)_4 \end{bmatrix}^{4+} \rightleftharpoons \begin{bmatrix} H_2O_4Cr \underset{O}{\overset{O}{\diagdown\diagup}} Cr(H_2O)_4 \end{bmatrix}^{2+} \qquad (2)$$

The order of the stability of chromium ion complexes with various ligands is presented below [8]:

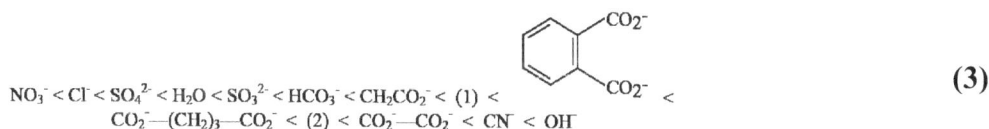

$$NO_3^- < Cl^- < SO_4^{2-} < H_2O < SO_3^{2-} < HCO_3^- < CH_2CO_2^- < (1) <$$
$$CO_2^- \text{—} (CH_2)_3 \text{—} CO_2^- < (2) < CO_2^- \text{—} CO_2^- < CN^- < OH^- \qquad (3)$$

with a benzene ring bearing two CO_2^- groups shown before the final $<$.

The stability of chromium-collagen complexes ranks between (1) and (2), which has a series of practical and theoretical implications.

The tanning process involves ligands present in chromium complexes which are replaced by the carboxyl groups of collagen.

Chromium-Collagen Compounds

There are several ways in which tanning action can take place, initially an ionic compound is formed, and then the most important chemical reaction certainly occurs through the covalent coordination bonds. Also it is certain that the most important ligands are carboxyl groups of the collagen macromolecule, as well as amino and imino groups of peptide bonds. This theory particularly explains the tanning action at various pH values. Tanning intensifies with increasing pH and does not occur when the collagen carboxyl groups are not ionized.

The factors influencing tanning are presented below:

Nature of anions. The tanning ability of chromium complexes increases depending on the nature of anions in the following order:

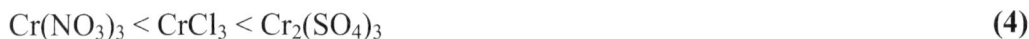

$$Cr(NO_3)_3 < CrCl_3 < Cr_2(SO_4)_3 \tag{4}$$

The anion associated with chromium salt is involved in bridge formation between chromium atoms; the SO_4^{2-} anion has the greatest ability to form these bridges (Fig. **8**).

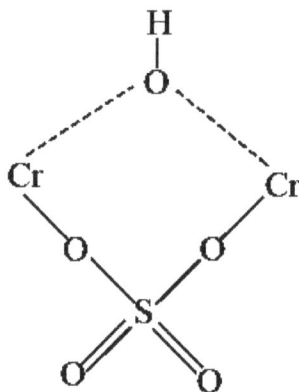

Figure 8: Bridges of sulfate with chromium ions.

Basicity of chromium salts. At a basicity of 30-40%, and pH ranging between 3.0-3.4, molecules with 2-4 chromium atoms prevail. This molecular size seems to be optimal for tanning; the length of these complexes coincides with the distance between collagen macromolecules.

At a lower basicity, the molecular size is too low and complexes do not manage to form bridges between collagen macromolecules, and at higher basicity values, molecules form semi-colloidal associations which cannot penetrate the fibrous collagen structure.

pH value. pH value of the solution is determined by the difference between the acidity of the solution and the conditions of balance. pH increase leads to intensification of tanning.

Complexing and masking agents. They are a major factor, influencing the type of leather which will be obtained. These complexing agents tend to reduce the tanning speed. In some cases, such as that of oxalates, if there are more than two molecules of oxalate to one chromium atom, the chromium complex will not tan.

From a practical perspective, the most important complexing agents are: sulphate, formate, acetate, phthalate, sulphite and dicarboxylic acids. Sulphate has low stability, even in the case of chromium sulphate in the form of powder. Formate is monodentate, it is frequently used and slows down the tanning process to such an extent that it leads to good quality leather. Acetate has an important effect on colour producing blue leathers. Phthalate is bidentate (occupies two coordination positions in the chromium complex), using this ligand leads to chromium exhaustion, freeing several positions for coordination of collagen groups.

Practical Aspects

In order to penetrate the fibrous structure of collagen, the molecules of the chromium complex must be small. This can be achieved by reducing the pH and the basicity of chromium salts. When the chromium salts have penetrated, the pH value must increase, which leads to ionisation of carboxyl groups of the collagen and formation of oLo and oxo polynuclear complexes. When tanning has taken place and the resistance of leather to temperature is higher, the bath temperature increases. Practically, tanning must take place in acid environment in order to have chromium penetration at low pH, followed by an increase of pH and finally, bath heating in order to obtain maximum chromium fixation.

Two examples are presented on next page.

Masking with Acetate

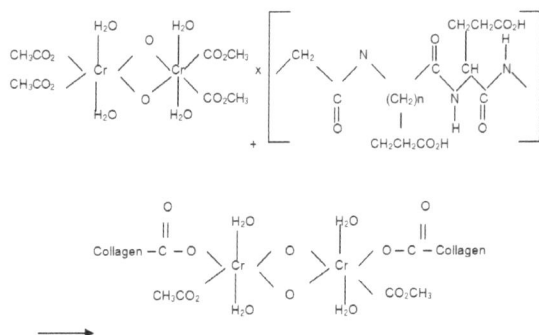

$$(5)$$

Leathers obtained with acetate masked chromium basic salt are even, supple, without a too high chromium content, coloured blue-green (suitable for white leathers). When acetate is used by itself, the link between chains occurs, the leather stays even and chromium content is not too high.

Masking with Phthalates

Leather obtained with phthalate masked chromium basic salt is fuller, not very even, has high chromium content and is green. Since phthalate is bidentate, chromium tends to exhaust better and several links are formed at the level of a macromolecular chain. The practical effect is a higher thermal stability.

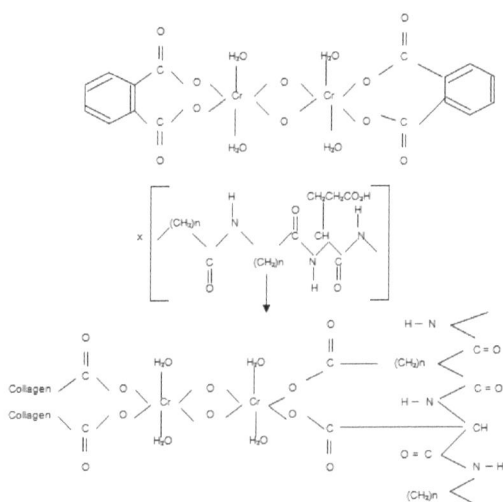

$$(6)$$

After completing any type of tanning, leather is tested in terms of thermal stability, in wet conditions. If it is resistant at 100°C, in boiled water, it is considered to be tanned sufficiently. This is the core of the tanning art and the main parameter for any challenging alternative for chromium tanning process.

CHAPTER 2

Chromium in the Leather Industry and its Environmental Implications

Abstract: Tanning using trivalent chromium salts underwent an evolution marked by the need to obtain technical performances for the leathers at first and then to limit the penetration of chromium in the environment by means of wastewater or solid waste. The evolution of methods to reduce environmental pollution with trivalent chromium salts included general environmental principles on advanced absorption of chromium, partial replacement of chromium, recycling, recovery and total replacement of chromium. Advanced technologies for exhaustion of chromium from effluents are based on the development of technological conditions favorable to absorption of chromium above the accepted level of concentrations by osmotic balance between the two environments, hide and tanning bath. Partial chromium replacement technologies are probably the most studied methods of reducing the use of chromium as tanning metal. The great diversity of variants proves that there is no imposed variant as of yet in the industry. Recycling and reuse of chromium are methods that require additional investment in facilities for recycling used tanning solutions or for precipitation and dissolution of chromium salts. The advantages consist in reducing environmental pollution and chromium consumption, while disadvantages are related to laborious analytical control and quality of leathers obtained, which is not at the same level as that obtained by using basic salts with a consistent quality. The newest way to reduce environmental pollution by chromium salts is wet-white tanning, which implies a pre-tanning with chromium-free materials, mechanical processing by splitting and shaving followed by tanning and retanning with chromium salts or syntans. The organic nature of chromium-free wet-white versions is the main advantage, but it also distances itself from the traditional, mineral character of natural leather. In this respect, the approach of tanning metallic hetercomplexes provides both classic features of mineral leathers and reduces environmental pollution.

Keywords: Advanced exhaustion, Chromium replacement, Chromium reuse, Chromium recovery, Wet-white technologies.

THE ENVIRONMENT AND THE USE OF CHROMIUM SALTS IN LEATHER TANNING

Since the discovery of chromium by Vauquelin in 1797 and to the moment of its application in tanneries, in 1893, it took nearly a hundred years of research and improvement of technologies for production and application of chromium tanning salts.

Chromium salts replaced potassium alum and vegetable tanning agents offering a number of major advantages which have granted them the status of universal material in the leather industry:

- much shorter duration of tanning;

- much higher hydrothermal resistance conferred to leather;

- the possibility of assortment diversification (versatility);

- much better tinctorial capacity than the chrome support compared to the vegetable one;

- better light resistance of leather;

- lower price.

Although much better than vegetable tanning, mineral tanning using chromium salts is not a panacea, it requires adjustments by retanning with chromium salts, vegetable tanning agents, synthetic tanning agents, synthetic resins, aldehydes *etc*.

The relatively uneven spread of chromium ores and their use in top strategic areas have been the first signs of concern that have stimulated research towards finding the first alternatives [11].

The second important moment in the evolution of chromium as a tanning material was the booming ecology that has impelled environmental legislation, largely restricting the uncontrolled expansion of chromium.

Thus, after the strategic factor, a new pressure starts to act on the leather industry - ecology.

The latter is becoming increasingly powerful acting towards the enactment of stringent limitations on water pollution, sludge and leather waste disposal.

The current environmental legislation demands that leather manufacturers:

- discharge wastewater containing chromium up to maximum 1.5-20 mg Cr^{3+}/dm^3 and 0.5-1 mg Cr^{6+}/dm^3 into municipal sewage [12,13];

- store sludge and leather waste in landfills for which large sums are paid or to treat sludge to a content of 1.2 g Cr/kg sludge for use in agriculture [14, 15];

- produce leathers with up to 3 mg/kg Cr^{6+} content [16, 17], or in which chromium is well fixed so that it would wash out only in a certain proportion (150 mg Cr/kg of leather) [18, 19].

The importance of "clean technologies" is recognized by all auxiliary materials and leather producers. All efforts in the field of tanning converge towards reducing or eliminating environmental pollution by chromium (III) salts – a heavy metal, which accumulates in plants and has the potential of turning into hexavalent chromium salts, which are even more toxic.

The regular tanning processes do not ensure compliance with the allowed limits for chromium concentration in wastewaters, not even in the case of mixing them with all wastewaters. Thus, conventional tanning processes lead to discharges containing 5-7g/ dm^3 Cr_2O_3 or even higher, which corresponds to chromium use yields of about 65%.

Therefore, efforts to improve chromium use yield, to recover used solutions, to find new tanning processes are the focus of research.

Currently, the following classification of methods to reduce pollution by chromium salts are accepted [20]:

- Methods of fixing-exhausting chromium;

- Partially replacing chromium with other tanning materials;

- Direct recycling of wastewaters;

- Recycling basic chromium salts from wastewaters;

- Using the wet-white tanning process.

METHODS OF FIXING-EXHAUSTING CHROMIUM FROM WASTEWATERS

Methods of fixing-exhausting chromium refer either to optimizing tanning operation parameters by increasing the temperature, pH, decreasing float volume, adding masking agents, ensuring optimal stirring, an optimal duration, or to the combination of these measures, finding materials with "catalytic" effect for chromium exhaustion, or using other solvents as reaction environments *etc.*

Temperature increase positively influences the tanning process up to a certain value.

The dependence of overall effective diffusion coefficient on temperature can be expressed by an Arrhenius equation:

$$D = D_0 e^{-Ea/RT} \qquad\qquad\qquad (7)$$

where: D_0- specific diffusion coefficient

Ea- activation energy for diffusion

R- ideal gas constant

T- absolute temperature.

Calculation of the activation energy for diffusion for a practical experiment [21] indicates the value of 18 kcal/mol, which confirms that the decisive stage of the tanning process is internal diffusion in leather dermis structure.

Given the mode of action of temperature on leather contraction and molecular size of the tanning solutions, in practice it is recommended to start tanning at 25°C and end it at 40-45°C.

Exceeding this temperature results in getting dark leathers, a coarser grain, low exhaustion efficiency, decreased tear strength, increased danger of precipitation of chromium complexes, which later affects dyeing quality.

The pH is probably one of the most important factors of adjusting the quality of the tanning process. It affects diffusion and binding of tanning complexes to leather. Thus, in the first tanning stage, the presence of non-ionized carboxyl groups and a chromium complex basicity of 33% is favored by a low pH, ranging from 2.5 to 2.8. Once the complex penetrates the leather section, a slow adjustment of pH leads to ionization of carboxyl groups which, in this form, are able to enter the chromium complex and increase the size of the chromium complex, which is thus able to bind to collagen chains and form cross-bridges, influencing hydrothermal resistance of collagen.

The relationship between reaction rate of chromium (III) and carboxyl groups of collagen is directly dependent on their degree of ionization and therefore on the pH [22].

Recent research indicates the possibility of obtaining very high tanning efficiency through basification at a relatively high pH [23].

Thus, through very slow basification it is possible to reach pH=6 without producing chromium precipitation and to achieve a 99.9% chromium exhaustion [24].

If the process is possible at pilot scale, an industrial process requires a level of process control that is not currently practiced.

The new generation of monitoring and control systems can, however, provide highly secure industrial technological processes which could not be imagined a long time ago [25].

Mechanical action is a very important parameter, especially for large leathers, with large cross-sections. Optimization of mechanical action is done while preserving leather quality. In essence, by increasing penetration rate of chromium in leather, the reactivity time increases, and thus, fixing time. The effect of mechanical action also depends on the float ratio and the size of hide batch.

Float ratio influences chromium penetration in the leather structure. The driving force of chromium penetration is determined by the tendency of equating the concentration in the float volume with that in the interfibrillary liquid.

In the case of pickling and tanning without float, there is little chromium tanning agent left in the residual float [26], but binding chromium tanning agent to dermal substance is poor because hydrolysis is not sufficient. In subsequent wet operations large amounts of chromium are washed out.

Cr_2O_3 offer directly influences the shrinkage temperature and is dependent on the reaction between carboxyl groups of leather and chromium.

Thus, the speed of reaction between chromium complex and leather is determined by the concentrations of chromium and carboxyl groups ionized according to the equation:

$$v = [\,Cr\,]^{\,a} \cdot [\text{-}CO_2]^{\,b} \tag{8}$$

In the case of a typical offer of 2% Cr_2O_3, for 1 kg of limed hide, which, in pickled state contains 0.1 mol carboxyl groups, the amount is 0.9-1.4 mol chromium [27].

Therefore, the speed of reaction will be favored by a higher content of chromium and ionized carboxyl groups.

Introduction of additional carboxyl groups in the hide leads to advanced chromium exhaustion.

Leather resistant to degradation can be obtained with an offer of only 0.7% Cr_2O_3, but the leather has a rough appearance. In practice an offer of 1.2-1.5% Cr_2O_3 is used, then a strong retanning is done, often using chromium or chromium-containing syntans. In this case, the problem of chromium in residual baths is not solved, it only becomes more difficult to deal with, because treatment of residual baths is further complicated by the presence of other retanning agents, dyes and fatliquoring agents.

Duration is another factor affecting tanning efficiency. Like any chemical reaction, tanning is a balancing reaction. In practice, extending the duration leads to an improvement of chromium exhaustion from the bath, not exceeding 65-80% of the chromium oxide offer [28].

The use of masking agents affect chromium penetration and allows basification at higher pH. Masking with monobasic or polybasic acids leads to changes in molecular size of the alkaline complex, changing electrical load and complexing ability.

A number of original Romanian research works have demonstrated the possibility of obtaining exhausted baths with 0.2-0.7 g/L Cr_2O_3 using dicarboxylic acids in various stages of tanning [29].

The commercial products of the most important basic chromium salts producer, Bayer AG, are organically masked salts ultimately leading to exhaustion yields of 95-98%. These procedures require strict working conditions, combining the principle of organic masking with traditional measures of chromium exhaustion in an original system called Baychrom C [30].

As variants of processes with masking agents, more recent research deal with other classes of carboxylic compounds such as acrylic acid polymers [31] or maleic acid-based polyelectrolytes [32].

Other tanning variants with advanced exhaustion implement the principle of creating new carboxyl groups active in collagen by initiating Mannich reaction [33].

Based on this principle, Hoechst company publishes a series of research results conducted with glyoxylic acid [34], dosed in the pickling operation.

Experiments performed at industrial level indicate the possibility of using an offer of 1-1.25% Cr_2O_3 and obtaining a bath exhaustion with 0.15-0.5 g/L Cr_2O_3, in the case of pickling bovine hides with 1.6% glyoxylic acid [35].

Achieving remarkable chromium exhaustion yields when using masking agents cannot be done without compromise: higher prices, the need to adjust operations of neutralization and dyeing, the final character of leather and difficulties in precipitating chromium during treatment [36].

Another series of interesting research is based on the idea that getting an advanced chromium exhaustion cannot be achieved under mild reaction conditions due to high affinity of chromium salts for the tanning environment -water.

Changing the reaction solvent leads to complete chromium salt exhaustion within minutes [37].

Problems that arise when tanning in nonaqueous solvents are the way of inducing chromium fixation in leather and the type of solvent used.

Other research works [37] propose using perchlorethylene as cost-effective solvent compared with the traditional method, given the complete regeneration of the solvent.

The existence of a European project to expand this process at industrial scale confirms the need for major changes in ecological leather tanning.

All these studies suggest that tanning with chromium salts, after a hundred years of application, still has unexplored possibilities of optimization.

PARTIALLY REPLACING CHROMIUM WITH ANOTHER TANNING MATERIAL

Research has shown that tanning with 0.5-0.8% Cr_2O_3 leads to chromium-free exhausted baths, but leather quality is poor even at an offer of 1% Cr_2O_3, therefore a good retanning is necessary to obtain leather of acceptable quality.

Previous attempts to replace chromium with basic aluminum or zirconium tanning agents showed that this replacement can be done to a certain extent.

The use of organic tanning agents from the class of synthetic tanning agents and aldehydes changes the character of leathers and physical-mechanical resistance, thus limiting the area of application.

Combining aluminum tanning agents with chromium (III) salts has led to an improvement of chromium exhaustion. A number of studies [38] have shown that the contribution of aluminum to chromium exhaustion is similar to the action of catalysts.

Rapid formation of complexes between aluminum and carboxyl groups of collagen causes a decrease in entropy of the system, ensuring a high reaction speed for chromium [27].

Combining chromium with other tanning metals, such as aluminum or titanium, allows to use a reduced Cr_2O_3 offer (1.25%) and to obtain an exhaustion yield of 98% [39].

Tanning using a mixture of aluminum, zirconium and chromium complexes was described many years ago and applied rather to get white, light resistant leathers. But these leathers too require a strong retanning using organic tanning agents in order to provide them with better fullness and softness.

It was found that by means of this tanning system very compact leathers are obtained which require postliming to reach a proper tanning depth.

Tanning metallic heterocomplexes are another group of materials that reduce chromium oxide offer and allow diversification of leather characteristics, and will be discussed in detail in a separate chapter.

Using glutaraldehyde before chromium tanning or after adding chromium in the tanning bath is already known from its introduction as tanning agent.

Glutaraldehyde was first introduced to improve the feel and fullness of leather, and later to replace chromium. It allows to reduce the chromium offer in tanning.

Using glutaraldehyde before chromium leads to fuller, softer leathers but with more open pores than if it were added after administering chromium. The process of chromium offer reduction by using glutaraldehyde was also the subject of recent research within a European research program [40].

It was found that an offer of 1.3% glutaraldehyde allows the use of only 1-1.5% Cr_2O_3 and the lower the Cr_2O_3 amount, the more the feel and mineral character of leather change. Thus, leathers become soft, spongy and are more difficult to dye.

These disadvantages can be eliminated by additionally using aluminum tanning agents.

Thus, using 0.5-1% glutaraldehyde, 1-1.5% Cr_2O_3 and maximum 1% Al_2O_3 in the form of aluminum chloride has led to good results by dosing tanning agents at every 1-2 hours. The process has not been tested at industrial scale.

Another series of procedures partially replace chromium oxide offer with zeolite-type aluminum silicate.

Dissolving zeolites at a pH lower than 5 leads to their hydrolysis and to the formation of basic aluminum salts and colloidal polysilicate acid with tanning effect.

The process leads to wastewater with 1g/l Cr_2O_3 if dicarboxylic acids are used in pickling and basification is done with magnesium oxide and aluminum silicate.

Also, a number of processes use formaldehyde or glutaraldehyde as pretanning agents, low chromium oxide offer and basification with aluminum silicate.

Other research [41] indicates the possibility of associating a reduced offer of chromium oxide of 1.25% with dosed ethanolamine in pickling and an exhaustion yield of 89-95%. Ethanolamine supposedly forms hydrogen bonds with carboxyl groups, thus contributing to the increase of shrinkage temperature of leathers.

Concerns in recent years [42-44] for modeling the molecular structure of collagen in order to predict the optimal structure of the ideal crosslinker and to understand the tanning process, indicate that replacing chromium with other tanning materials is still the subject of current research.

DIRECT RECIRCULATION OF WASTEWATERS

This method was applied in Japan [45], in some tanneries in England and for sheep skins in Australia [46].

Used tanning baths must be first subjected to filtration treatments to separate solid impurities and even fat.

Due to the fact that tanning can be done either in pickling baths, or in new baths, there are 2 variants of recirculation:

1)   Recycling pickling baths in which tanning was carried out

Filtered exhausted baths are combined with 1/2 of the usual pickling acid, after 10 minutes stirring with pelt, the second half diluted with residual bath is added. The density is checked, and after 1-2 hours of pickling, the usual chromium tanning agent is added, of which the amount of chromium oxide already introduced through the residual bath is subtracted.

2)   Reusing residual chromium baths from the tanning bath

Normally pickled hides are tanned in the residual chromium bath of the earlier stage to which fresh tanning agent is added, given the existing amount in the residual bath.

The basifier is reported only to the quantity of fresh tanning agent. This method can save 10-33% of Cr_2O_3 offer.

Bath reuse also leads to a reduction of Cr_2O_3 load in residual baths, thus reducing treatment costs.

Reservations in applying this process refer to a number of disadvantages:

- increase in volume of exhausted baths which cannot be recycled [34];

- the possibility of accumulation of impurities can affect the skin quality;

- the need for a strict analytical control of baths;

- the possibility of obtaining differently coloured leathers, usually darker.

More recent research on the possibility of recycling residual baths from high-exhaustion tanning indicates that in this case, the method is not economical, providing only 2% of chromium oxide offer [40].

RECYCLING BASIC CHROMIUM SALTS FROM WASTEWATERS

Chromium salts from exhausted baths may undergo alkaline precipitation, separation by decantation, filtration, redissolving in acid and then reuse.

Depending on the choice of precipitating agent, coagulation and separation parameters are different. Thus, three variants can be identified:

1) fast precipitation with sodium hydroxide or sodium carbonate, creating a voluminous precipitate which is separated by filtration or centrifugation;

2) fast precipitation using polyelectrolytes and dehydration of precipitate by pressing;

3) slow precipitation with magnesium oxide which produces a relatively dense precipitate that can be decanted.

Of course, the most economical method is the slow precipitation.

In practice, both chromium from chroming baths and from hide wringing may be precipitated.

In theory, washing baths and retanning baths can be treated in the same way. However, due to the low concentration of chromium in these baths and to the substances accompanying these solutions, this process cannot be applied.

The method is advantageous due to the high degree of chromium removal from wastewater, so that it can be compared in terms of ecological efficiency to Baychrom C procedures [47, 48].

Reservations on the application of this process are related to the following aspects:

- the need for rigorous analytical control;

- consumption of chemicals and personnel expenses, which reduce economic efficiency;

- complex systems that require a large capital;

- lighter color of leather.

At present, concerns for the application of modern methods for regenerative separation of chromium salts and for their recovery is heading towards processes using ion exchange [49], ultrafiltration membranes [50], air flotation [51] *etc.*

USING THE WET-WHITE TANNING PROCESS

In recent years a new tanning system and a new ecologic concept -clean leather - emerged, the wet-white or "blanc stabilisé humide", as French researchers called it.

This new technology requires pretanning leathers with agents able to stabilize the skin up to shrinkage temperatures of about 68-74°C, in order to withstand the mechanical operations of equalization.

The remarkable advantages of this method are:

- obtaining chromium-free exhausted baths;

- obtaining chromium-free leather waste that can be easily recovered;

- the possibility of storing and marketing wet-white leather;

- reducing the cost of treatment.

Although these technologies have made their mark on the industry only in a very limited area, namely leather for automotive upholstery, they are considered to be an acceptable alternative when banning the production of chrome leather waste [48].

It is noted that the largest chemical companies and most prestigious researchers have developed wet-white technologies in recent years.

Thus Professor Heidemann indicates different types of achieving pretanning [52] with vegetable tanning agents with syntans.

In 1990, L.Tonigold and E. Heidemann were carrying out studies on the pretanning and retanning ability of iron [53]. The conclusions are that iron (III) is in some ways superior to other tanning metals, contrary to previous research findings.

Thus, a chromium-free leather version is done, called wet-brown or wet-iron, with shrinkage temperature of 70°C.

M. Siegler, one of the initiators of chromium-free tanning technologies [54] carried out, in collaboration with Rohm & Haas company (USA), a white tanning using a complex mixture consisting of a polyaldehyde, a tertiary amine and a monohydric alcohol.

In 1991, the ICI company presents a heterocomplex based on Al, Ti and Mg, the result of an investigation begun in 1984, which is able to completely replace chromium [55, 56].

This product, called Synektal TAL, was used as retanning agent in tanneries in England.

R. Celades [57] develops various chromium-free leather assortments using a titanium salt and a polysaccharide with the role of increasing the number of functional groups of collagen.

In France, G. Gavend, C. Rabbia and J.P. Communal (Rhône-Poulenc Chimie) develop a product based on organically complexed aluminum called Rhoditan, which is able to give leather hydrothermal resistance of 68°C [58].

In 1992 Ciba Geigy company presents at the Paris Leather Week a wet-white technology [59] based on which ten leather assortments, from vegetable leathers, nubuck, upholstery leather, to sheep furs.

Pretanning is done using a modified glutaraldehyde-based product, which, together with the pickling tanning agent, leads to a leather with shrinkage temperature of 72-74°C.

In Italy, U. Sammarco [60] develops wet-white leathers using aluminum-aldehyde based compounds. Shaved leathers are then chromium tanned in a system of almost total absorption. He obtains leathers with hydrothermal resistance of 80°C.

In 1993 [61], J. Pore and Hoechst company specialists developed an assortment of wet-white leather using a colloidal silica based product (Filiderm W).

A number of Romanian original research works [62, 63] present a chromium-free tanning product based on aluminum, zirconium and magnesium, by means of which wet-white leather can be obtained [64-71].

Research on salts of zirconium, iron and rare metals demonstrates that there are insufficiently explored paths in this direction. The zirconium 4-hydroxymandelate complex [72] crosslinks collagen up to a shrinkage temperature of 97°C, compared to another polymeric zirconium complex, Organozir [73] which crosslinks collagen at 90°C, under the conditions of a stability to alkali of up to a pH value of 5. Rare metals from the cerium family and from mixtures of rare earths [74] show potential to be used alone or in combination with chromium salts, with economical and ecological advantages compared to using only chromium salts.

In the area of wet-white organic tanning, many developments have shown the possibility of using oxazolidine [75], tetrakis hydroxymethyl phosphonate (THPS) in various combinations [76]. The disadvantages of using these materials were the higher price, the need to use them in combination with other materials that raise costs or increase pollution.

Very recently [77] new commercial products were launched for semi-organic or organic wet-white pretanning of leathers using a combination of aluminum and silicone (Tanfor T, Kemira, 2012), polymeric aldehyde (Cromogenia, 2012), poly carbamoyl sulfonate (PCMS, Lanxsess, 2011) *etc.* The great product diversity shows the existence of a market demand for alternatives to the use of basic chromium salts and the beginning of a new era in natural leather tanning, where chromium will no longer have a 80-90% share of world production.

We conclude that at present, environmental concerns have spurred research towards finding efficient ways to optimize the use of chromium salts, while maintaining leather quality. Thus, methods for reducing the chromium offer and its advanced exhaustion from residual baths play a very important role.

Maintaining the unique characteristics of mineral tanned leathers, unrivaled by any organic compound, will remain an important goal of producers and consumers of natural leather.

This direction comprises the use of tanning metallic heterocomplexes, the subject of this work.

CHAPTER 3

Synthesis and Use of Tanning Metallic Heterocomplexes

Abstract: Developing complex metallic heterocomplexes requires knowledge of synthesis methods based on oxidation-reduction reactions, complexation to boiling point or basification of mixtures of salts of chromium, aluminum, iron, zirconium or titanium, the main tanning metals for natural leather. The purpose of these syntheses is developing heterocomplexes with higher stability to alkali than that of metal salts from which the natural leather tanning properties derive. Chromium-aluminum salts are the only heterocomplexed salts that are currently commercially available, with restricted use to certain types of applications related to maintaining a lighter colour of leather and improved polishing ability. Knowing the possible structure of complex metallic heterocomplexes obtained is a complicated endeavor due to the polydispersity of tanning solutions and the difficulty of separating the tanning heterocomplex in crystal form. IR analyses enabled the formulation of hypotheses on the structure of chromium-zirconium and aluminum-titanium heterocomplexes, while X-ray diffraction of chrome-zirconium-cerium crystal enabled the identification of the role of sulfate ion of ligand tridentate and not bidentate, as it was previously thought, the identification of oximetallic O-Cr-O bond length of 10Å, identical to the distance between the collagen macromolecules forming the three-macromolecule helix. The close connection between the synthesis method and the stability of tanning metallic heterocomplexes is highlighted by examples of research in the field. The topicality of tanning metallic heterocomplexes is emphasized by their superior properties compared to organic tanning variants, both in terms of economy and technology, and the greater possibility of combining metals or organic ligands to develop structures with different collagen crosslinking properties.

Keywords: Synthesis methods, Chromium-aluminum tanning agents, Tanning heterocomplexes, Structure of heterocomplexes, Properties of heterocomplexes.

SYNTHESIS, STRUCTURE AND PROPERTIES OF TANNING METALLIC HETEROCOMPLEXES

Tanning metallic heterocomplexes are complex polynuclear combinations containing at least two different complex generators: $Cr(III)$, $Al(III)$, $Fe(III)$, $Zr(IV)$, $Ti(IV)$ bound together by various ligands such as: OH^-, SO_4^{2-}, SO_3^{2-}, $HCOO^-$, $C_2O_4^{2-}$, $CH_3\text{-}COO^-$ *etc*.

Developing heterocomplexes for leather tanning is of great importance because complexation leads to a higher pH value, at which the precipitation of these mixed complexes takes place, compared to isopolycomplex tanning solutions of Al(III), Fe(III), Zr(IV) and Ti(IV).

Developing tanning metallic heterocomplexes comprising at least two different complex generators is done by various patented methods. Existing data in the literature outline three main directions for the preparation of these products:

- oxidation-reduction methods that require reducing hexavalent chromium to trivalent chromium in acid medium and in the presence of another tanning metal salt using, as appropriate, organic or inorganic reducing agents;

- methods that require heating up to boiling point of solutions of various metal salts in the presence of organic or inorganic complexing agents;

- methods that require the cold or hot use of agents with complexing and basification effect of different solutions of different tanning metal salts.

Chromium-aluminum heterocomplexes were manufactured at industrial scale by reducing sodium dichromate in the presence of aluminum sulfate and sulfuric acid using glucose as a reductant [78-80].

The general reaction for the preparation of chromium-aluminum heterocomplexes developed in this research is the following:

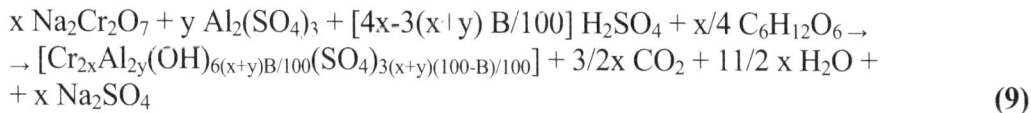

$$x \ Na_2Cr_2O_7 + y \ Al_2(SO_4)_3 + [4x-3(x+y) \ B/100] \ H_2SO_4 + x/4 \ C_6H_{12}O_6 \rightarrow$$
$$\rightarrow [Cr_{2x}Al_{2y}(OH)_{6(x+y)B/100}(SO_4)_{3(x+y)(100-B)/100}] + 3/2x \ CO_2 + 11/2 \ x \ H_2O +$$
$$+ x \ Na_2SO_4 \tag{9}$$

where:

$$x=0.329(1....1.98);$$

$$y=0.049(10....0.20);$$

$$B=basicity$$

With this formula the whole range of chromium-aluminum heterocomplexes can be prepared with variable proportions of Cr_2O_3 and Al_2O_3, from 99/1 to 50/50.

For aluminum-chromium heterocomplexes, namely those containing more aluminum than chromium, the general reaction for preparation developed is as follows:

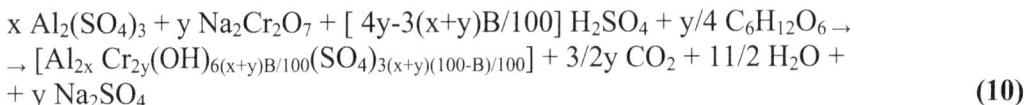

$$x\ Al_2(SO_4)_3 + y\ Na_2Cr_2O_7 + [\ 4y-3(x+y)B/100]\ H_2SO_4 + y/4\ C_6H_{12}O_6 \rightarrow$$
$$\rightarrow [Al_{2x}\ Cr_{2y}(OH)_{6(x+y)B/100}(SO_4)_{3(x+y)(100-B)/100}] + 3/2y\ CO_2 + 11/2\ H_2O +$$
$$+ y\ Na_2SO_4 \tag{10}$$

where:

$$x=0.049(19.8\ldots\ldots10);$$

$$y=0.329(0.02\ldots\ldots.1);$$

$$B=basicity$$

This formula is applicable for manufacturing all aluminum-chromium heterocomplexes containing relations between Al_2O_3 and Cr_2O_3 from 99/1 to 50/50, *i.e.* Al_2O_3 content is higher than or equal to that of Cr_2O_3.

In the formation of heterocomplexes containing chromium and aluminum, a particularly important issue is establishing the quantity of sulfuric acid to be added to perform the synthesis, given that aluminum sulfate is a generator of sulfuric acid during the reaction.

Under these conditions, based on theoretical calculations and practical experiments, a general formula for determining the amount of sulfuric acid required to be added to the synthesis of these tanning heterocomplexes was elaborated, taking into account both the Cr_2O_3/Al_2O_3 ratio and the desired basicity.

This formula is shown below:

$$A=258.1.C\ Cr_2O_3/100 -[193.5.C\ Cr_2O_3/100 + 288.6.C\ Al_2O_3/100].\ B/100 \tag{11}$$

where: A = amount of 100% necessary sulfuric acid

C Cr_2O_3 = concentration of Cr_2O_3

C Al_2O_3 = concentration of Al_2O_3

B = desired basicity.

The oxidation-reduction can also lead to developing heterocomplexes stabilized with organic ligands such as formic acid, lactic acid, tartaric acid, citric acid or syntans [81, 82].

To obtain such a chromium-aluminum heterocomplex, the following synthesis reaction is used:

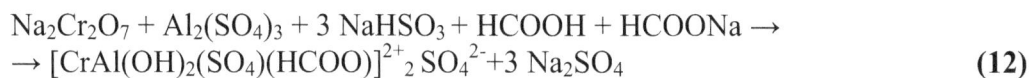

$$Na_2Cr_2O_7 + Al_2(SO_4)_3 + 3\ NaHSO_3 + HCOOH + HCOONa \rightarrow$$
$$\rightarrow [CrAl(OH)_2(SO_4)(HCOO)]^{2+}_2\ SO_4^{2-} + 3\ Na_2SO_4 \qquad\qquad (12)$$

The reaction is carried out in highly concentrated solutions and masking agents are added after total conversion of hexavalent chromium to trivalent chromium.

The average formula of chromium-aluminum heterocomplexes formed using this method is considered to be the following (Fig. **9**).

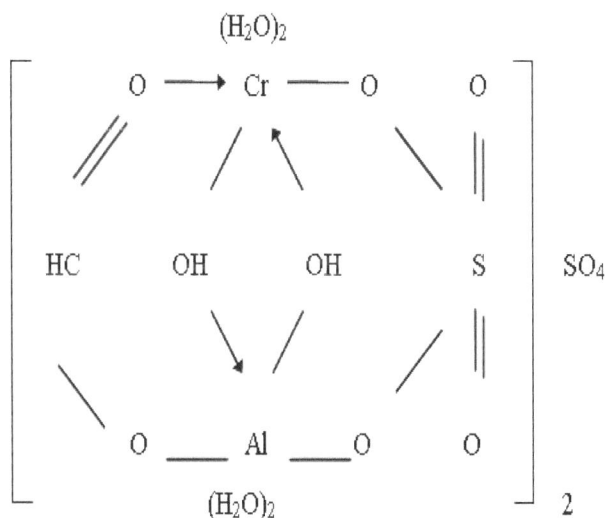

Figure 9: Di-μ-formato-di-μ- hydroxo -μ-sulfato- tetraaqua Al(III) Cr(III) sulfate.

The chemical characteristics of chromium-aluminum solution complexed with organic ligands are shown in Table **3**.

Table 3: Characteristics of the chromium-aluminum solution

Characteristics	MU	Value
Concentration of Cr_2O_3	%	11.1
Concentration of Al_2O_3	%	7.1
Basicity	%	40.5
Concentration of SO_4	%	21.5

The use of the product was either dry (without bath) or by spraying and resistance to boiling temperature was obtained only after neutralization.

In 1967, BASF company releases a patent [83] which describes a trinuclear Cr-Al complex with the following formula:

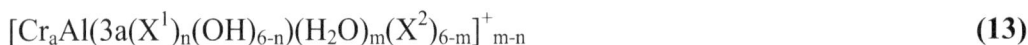

$$[Cr_aAl(3a(X^1)_n(OH)_{6-n})(H_2O)_m(X^2)_{6-m}]^+_{m-n} \qquad (13)$$

where: X^1, X^2-acyloxy aliphatic radicals with monocarboxyl C_{1-3} ;

X^1 is combined with 2 central atoms;

X^2 is combined with one central atom; a = 1 or 2; n = 2 or m = 4-6.

Another patent [84] describes developing a trinuclear chromium-aluminum heterocomplex by complexing a mixture of basic chromium salts and aluminum with acrylic acid. The structure proposed by the authors for the developed complex is shown below (Fig. **10**).

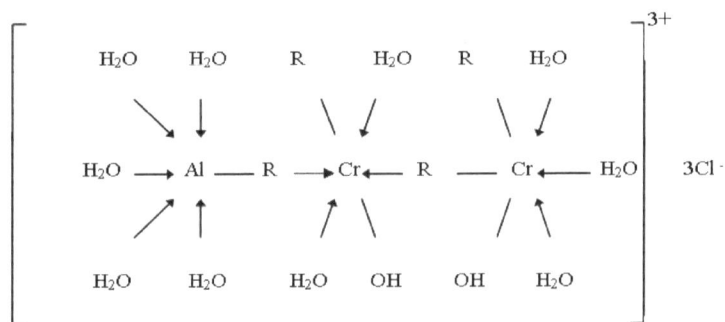

Figure 10: Di- μ-acrylato-diacido-dihydroxo-decaaqua-monoaluminum(III)-dichromium (III) chloride.

where: R-radical of acrylic acid

Tanning hides using this complex leads to full and very supple leathers with shrinkage temperature of 78-80°C.

Synthesis of chromium and aluminum heterocomplexes by reducing Cr(VI) in aqueous solutions in the presence of $Al_2(SO_4)_3$ is described in 1989 by B.S. Shimenovide and Mihailov [85].

They develop chromium-aluminum complexes, with 2-4 central atoms, whose polynuclearity is obtained through ligands OH[-] and SO_4^{2-}.

Another study conducted in 1990 in Spain [86] highlights the qualities of leathers tanned with chromium-aluminum salts, compared to those of leathers tanned with chromium salts. The development, the properties of chromium-aluminum complexes and the characteristics of leathers tanned using these complexes are also presented by Russian authors in a number of publications in 1991 [87,88].

Cr-Al solutions are typically turned into powder by various methods (atomization, cylinder drying) and marketed under different names: Lutan CR, Blancorol AC, Securcrom 30 Al *etc*.

The main characteristics of commercial chromium-aluminum heterocomplexes are given in Table 4 [89-91].

Table 4: Commercial chromium-aluminum heterocomplexes and their characteristics

Name	Company	Content of			Basicity	Recommendation of use
		Al_2O_3 (%)	Cr_2O_3 (%)	Na_2SO_4 (%)	(%Sch)	
Blancorol AC	Bayer AG	14	7	7	50	Retanning and tanning white leathers and suede leather.
Lutan CR	BASF	14	7	-	50	Tanning white leathers and sheep furskins
Retanal AC	Cromogenia UNITS	8	5	-	-	Retanning white leathers

In general, chromium-aluminum heterocomplexes are recommended for white and pastel leathers, suede leather (improved polishing ability), fur with white wool (tips of wool do not turn green) and bright dyes. Literature also reports development of chromium-iron heterocomplexes with the following structures (Fig. **11, 12**).

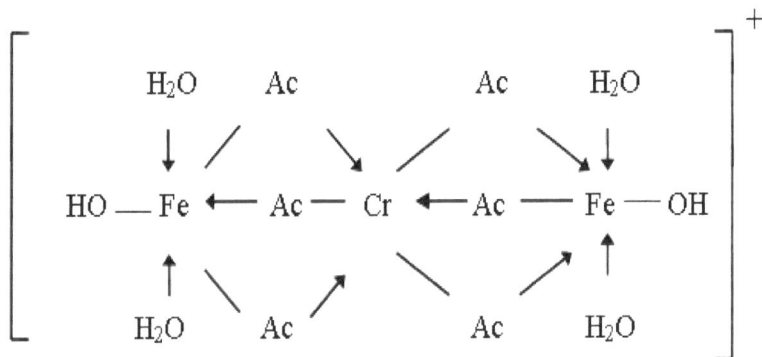

$$[Fe_2Cr(OH)_2(CH_3COO)_6]^+ \tag{14}$$

Figure 11: Complex cation of hexa-μ-acetato-dihydroxo-tetraaqua-diiron(III) chromium(III) (Ac=Acetic acid).

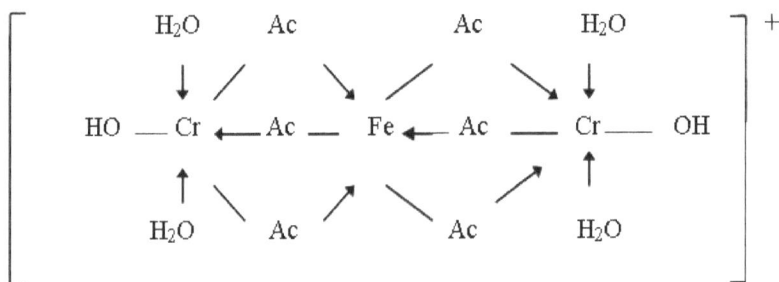

$$[FeCr_2(OH)_2(CH_3COO)_6]^+ \tag{15}$$

Figure 12: Complex cation of hexa-μ-acetato-dihydroxo-tetraaqua-dichromium(III) iron(III) (Ac=Acetic acid).

Using the method of heating up to boiling point of tanning metal sulfates in the presence of sodium and ammonium sulfate, the chromium-zirconium heterocomplexes were obtained with a similar structure as shown in Fig. **13**.

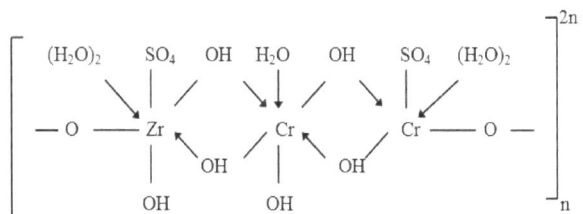

Figure 13: Complex anion of tetra-μ-hydroxo-dihydroxo-di-μ-oxo-dysulfato-tetraaqua-dichromium(III) zirconium(IV).

The first complete structural analysis of a mixed complex identified in literature, the chromium-zirconium complex [92], obtained by reduction of sodium dichromate mixture and zirconium sulfate with sulfur dioxide showed the following compound:

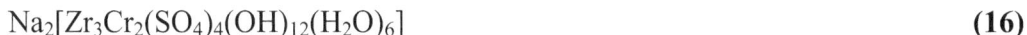

$$Na_2[Zr_3Cr_2(SO_4)_4(OH)_{12}(H_2O)_6] \tag{16}$$

Chromium-zirconium heterocomplex solution stability lies around pH = 3.3, higher than that of basic solution of zirconium sulfate (precipitation pH is 2.3).

Isolation and characterization of three-dimensional structure of chromium-zirconium heterocomplex involved using sophisticated techniques for obtaining the monocrystal by replacing sodium with cesium and then analysis by X-ray diffraction [92].

The three-dimensional structure determined by this technique is shown in Fig. **14**.

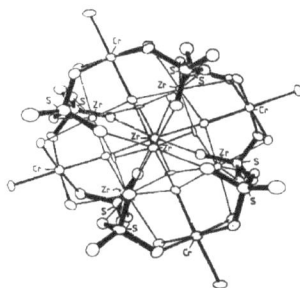

Figure 14: Three-dimensional structure of a chromium-zirconium heterocomplex [92].

Such structural analysis was able to very precisely define the isolated product and revealed possibilities of reaction with the hide. Also, the role of SO_4^{2-} group of tridentate ligand, and not bidentate, as thought in previous investigations, was found; the polyoxymetal group was identified, where the distance between the O-Cr-O atoms is about 10 Å - the ideal distance for cross-bridge formation in the collagen structure. It can be therefore assumed that such structural elements may be essential for the qualities of a new tanning heterocomplex.

Chromium-zirconium heterocomplexes with higher stabilities are obtained when hydroxycarboxylic acids are added to the complexes.

Also, the ratio of chromium and zirconium leads to the formation of complexes with different loads and different properties. Thus, increasing the proportion of zirconium atoms, predominantly anionic complexes are obtained.

Zirconium containing heterocomplexes give leather a full up to firm character.

Electron microscopy studies [93] have shown that, unlike basic zirconium salts that are deposited between fibers, similar to vegetable tanning agents, chromium-zirconium heterocomplexes do not deposit, but bond to the fiber.

Many research studies have been conducted by manufacturers of titanium oxides (Tioxide Group PLC) in order to extend the use of this resource in leather tanning.

Thus, a number of patents describe various methods of synthesis of aluminum-titanium heterocomplexes by stabilization with hydroxycarbonate acids [94] or magnesium oxide [95-111].

Table 5: Characteristics of aluminum-titanium-magnesium heterocomplex in the form of powder

Characteristics	MU	Value
Appearance		White powder
Density		0.8
pH solution 10 g/l		3.9
Dry substance	%	95
Metal oxides	%	22

TAL Synektal product of the ICI company is an aluminum-titanium-magnesium heterocomplex. It allows reducing the chromium offer in tanning or retanning, obtaining chromium-free wet-white leather and has the features presented in Table **5**.

A French patent [112] describes a process for tanning leather using complexes of aluminum(III) and titanium(IV) in combination with a synthetic tanning agent.

Other research studies [113] highlight the advantages of tanning sheep furskins using complexes of aluminum(III) and titanium(IV) compared to chromium salt tanning. Aluminum-zirconium complexes were investigated in combination with various syntans to determine optimal use conditions [114].

Obtaining assortments of wet-blue leathers with low-chromium content was studied by using combinations of chromium-aluminum-zirconium allowing substantial reduction of wastewater pollution. The study [114] focuses on great possibilities for improvement of physical and mechanical characteristics of leathers by optimizing retanning-fatliquoring formulation.

Tanning and retanning with titanium-aluminum complexes is investigated also in terms of the possibilities of reducing the use of auxiliary chemicals [99]. Another economic alternative of using aluminum-titanium complexes is for tanning leather intended for soles to reduce consumption of vegetable tanning agents [106, 107].

Other patents [108] describe the development of mineral aluminum-zirconium-titanium tanning agents with different compositions and ratios of metal oxides: $Al_2O_3 : (ZrO_2 + TiO_2)$ of 0.5-2 : 1n or $ZrO_2 : TiO_2$ of 0.5-2 : 1.

Another method of obtaining Zr-Al-Cr heterocomplexes considered advantageous is described by a Bayer patent [109], in which complexation is achieved by heating-dehydrating aqueous solutions by spraying or cylinder drying.

Tanning using mineral complexes based on iron and aluminum is also approached by Japanese researchers [110]. Interactions that occur in tanning aqueous solutions of salts of iron, zirconium, titanium, aluminum are studied by Russian scientists [111]. Studies conducted by professor Mandrev on heteropolynuclear chromium-zirconium, chromium-titanium and chromium-titanium-zirconium complexes lead to the conclusion that the role of chromium in their structure is that of a stabilizer. Development of a series of aluminum complexes, such as: aluminum-zirconium, aluminum-titanium, aluminum-titanium-zirconium, with stability ranging from pH 3.3 to 3.7 and structural studies carried out revealed the role of aluminum in the stability of complexes. Thus, with the increase of aluminum share in the complex, an elimination of bidentate SO_4 groups and their transfer outside the coordination sphere occur. It is obvious, therefore, how the complex type is influenced by the extent and type of tanning metals in its composition.

Based on a series of structural investigations of separation and IR analysis, professor Mandrev [112] issues the hypothesis of formation of the following structures (Figs. **15, 16**).

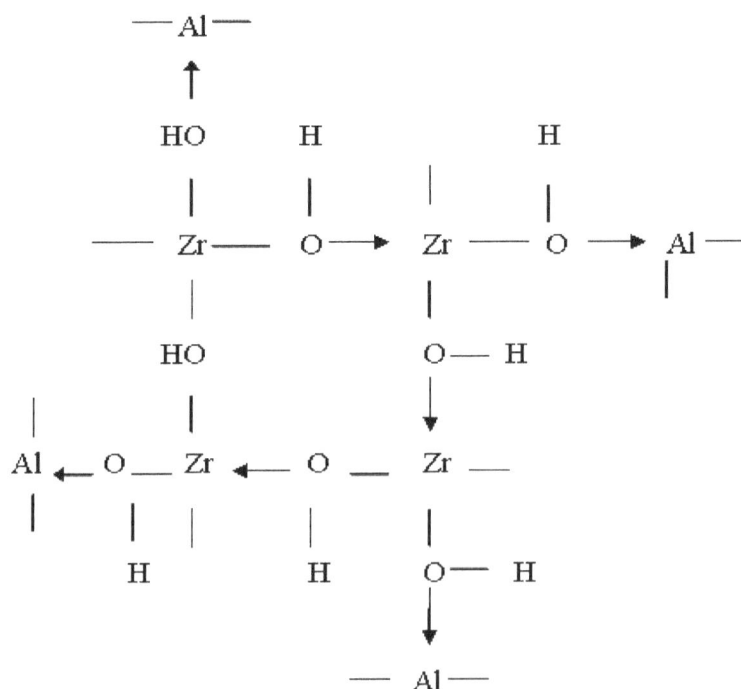

Figure 15: Octa-μ-hydroxo-tetrazirconium(IV)tetraaluminum(III) chain.

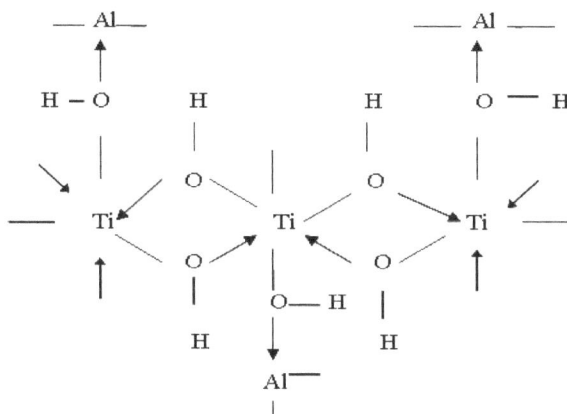

Figure 16: Hepta-μ-hydroxo-trititanium(IV)trialuminum(III) chain.

Studies on stabilization of iron salts by interaction with other tanning metals and polynuclear heterocomplexes formation were discussed by Russian researchers [113]. The authors manage to develop a series of mixed complexes of Cr-Fe, Cr-Fe-Zr, Cr-Fe-Al and Cr-Fe-Ti with stabilities ranging from 2.7 to 3.1 pH values, highlighting the stabilizing role of zirconium and launching the hypothesis of a structure of Cr-Fe-Zr complex as follows (Fig. **17**).

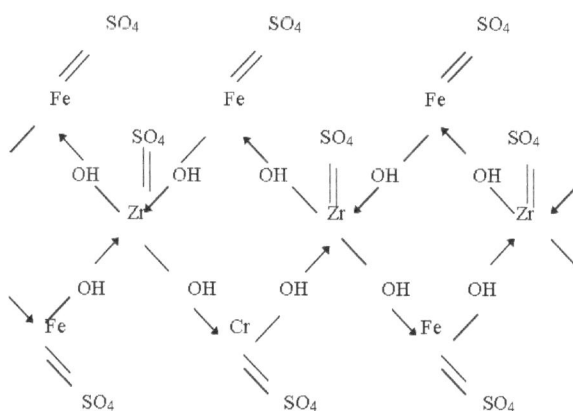

Figure 17: Deca-μ-hydroxo-nona-sulfato-trizirconium(IV)chromium(III)pentairon(III) chain.

The low stability of developed heterocomplexes is due to the method of synthesis which involved the use of the oxidation-reduction reaction only for reduction properties of ferrous sulphate in the presence of sodium sulfo zirconate as complexing agent. Also, drying the product thus obtained by calcination resulted in an additional anionization of complexes formed and in a slight improvement in alkali stability.

From the technological point of view, it is currently accepted that tanning metallic heterocomplexes are characterized by:

- high penetration rate;

- easy application;

- the possibility of obtaining leathers with appropriate shrinkage temperatures;

- the possibility of obtaining leathers similar to chrome leathers;

- the possibility of obtaining exhausted baths with low chromium content or chromium-free;

- the possibility of reducing chromium consumption;

- the possibility of reducing pollution-related costs;

- the possibility of obtaining leathers with enhanced features: fullness, more intense color, polishing ability *etc.*

Returning to variants of leather tanning using metal salts alternative to basic chromium salts is due to the higher cost of organic matter alternatives, the much different properties from those of mineral tanning and induction of new pollutants in effluents. Chromium salts manufacturers [114] have selected iron as metal without ecological implications and have developed chromium-free tanning technologies, concluding that the disadvantages of its use are lower than using organic crosslinkers.

Tanning metallic heterocomplexes are one of the viable alternatives because the diversity of their structures and their synergism of action offers the possibility of multifunctional collagen crosslinking, a requirement to compete with the performance of chromium as tanning agent [115].

Send Orders for Reprints on reprints@benthamscience.net

CHAPTER 4

Theory and Experimentation of Synthesis Reactions of Tanning Metallic Heterocomplexes

Abstract: The main synthesis reaction of metallic heterocomplexes is the oxidation-reduction reaction in acid medium using glucose as a reducing agent. Understanding the mechanism of synthesis of metallic heterocomplexes with various basicities is related to the stoichiometric mechanism of sulfuric acid release by the aluminum sulphate present in the system, to the use of sulfuric acid by the ferrous sulfate which oxidizes to ferric sulfate. The theoretical study of the synthesis reactions of tanning heterocomplexes reviews all the basicity values which can be obtained and proposes formulas for calculating the amount of sulfuric acid required to obtain a quantity of tanning metallic heterocomplexes containing 100 kg of metal oxides. Monitoring the laboratory scale synthesis of chromium-iron, chromium-aluminum-iron and iron-chromium-zirconium metallic heterocomplexes confirms the stoichiometric mechanism for heterocomplexes with 33% basicity. The influence of aluminum and iron in metallic heterocomplexes consists in destabilizing the complex so that the chromium-iron and the chromium-iron-zirconium heterocomplexes are more stable than the chromium-aluminum-iron one. Stability to alkali of the new tanning metallic complexes confirms internal heterocomplexation and the formation of more stable structures than the monometallic salts they derive from. Stability of the three types of tanning metallic heterocomplexes over time is good after 20 days up to 5 months and demonstrates the possibility of using them in leather tanning and retanning under similar conditions to using basic chromium salts. Tanning metallic heterocomplexes enable Cr_2O_3 offer reduction by 30-50% and therefore reduction of pollution provided that a synergy of interaction with collagen is achieved. Pilot scale synthesis of chromium-iron, chromium-aluminum-iron and chromium-iron-zirconium variants of heterocomplexes with the best stability has validated the theoretic reactions developed and showed the higher stability of tanning metallic heterocomplecxes. Tanning metallic heterocomplexes can also be developed as powder by atomization, and the stability to alkali of solutions obtained from powders is higher in the case of chromium-iron heterocomplexes and remains unchanged for other types of tanning heterocomplexes.

Keywords: Synthesis of heterocomplexes, Basic chromium salts, Chromium-aluminium heterocomplexes, Chromium-iron heterocomplexes, Chromium-aluminium-iron heterocomplexes, Chromium-zirconium heterocomplexes, Chromium-aluminium-zirconium heterocomplexes, Chromium-iron-zirconium heterocomplexes.

Carmen Gaidau

THEORETICAL SYNTHESIS REACTIONS OF TANNING METALLIC HETEROCOMPLEXES

Tanning metallic heterocomplexes can be obtained using the classical redox reaction in sulfuric acid medium and in the presence of glucose as a reductant, used in the synthesis of chromium complexes.

The existence of a reaction system, in which several types of metals, such as chromium, aluminum, iron, zirconium or titanium with different properties are found, required the development of a new concept of the synthesis mechanism to underlie experiments and studies on the developement and application of new tanning metallic heterocomplexes.

Thus, the presence of several sulfuric acid generators or of more reductors in the system requires a study on the theoretical possibilities of conducting synthesis so as to obtain complexes with the best composition and tanning capacity.

To understand the synthesis mechanism of various tanning metallic heterocomplexes, theoretical formation reactions of complexes of chromium, chromium-aluminum, chromium-iron, chromium-aluminum-iron, chromium-zirconium, chromium-aluminum-zirconium and chromium-iron-zirconium were studied depending on the amount of sulfuric acid in the system and the basicty variation.

For this purpose, sodium dichromate, aluminum sulphate, ferrous sulphate and zirconium sulfate, sulfuric acid as reaction medium and glucose as a reductant were used.

The synthesis reaction of chromium complexes depending on the amount of sulfuric acid and basicity variation of the obtained complexes is given below as reference for tanning metallic heterocomplexes:

Basic Chromium Complexes

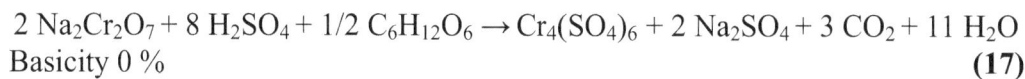

$2\ Na_2Cr_2O_7 + 8\ H_2SO_4 + 1/2\ C_6H_{12}O_6 \rightarrow Cr_4(SO_4)_6 + 2\ Na_2SO_4 + 3\ CO_2 + 11\ H_2O$
Basicity 0 % **(17)**

$$2\,Na_2Cr_2O_7 + 7\,H_2SO_4 + 1/2\,C_6H_{12}O_6 \rightarrow Cr_4(OH)_2(SO_4)_5 + 2\,Na_2SO_4 + 3\,CO_2 + 9\,H_2O \text{ Basicity } 16.6\,\%$$
(18)

$$2\,Na_2Cr_2O_7 + 6\,H_2SO_4 + 1/2\,C_6H_{12}O_6 \rightarrow Cr_4(OH)_4(SO_4)_4 + Na_2SO_4 + 3\,CO_2 + 7\,H_2O \text{ Basicity } 33.3\%$$
(19)

$$2\,Na_2Cr_2O_7 + 5\,H_2SO_4 + 1/2\,C_6H_{12}O_6 \rightarrow Cr_4(OH)_6(SO_4)_3 + Na_2SO_4 + 3\,CO_2 + 5\,H_2O \text{ Basicity } 50\,\%$$
(20)

$$2\,Na_2Cr_2O_7 + 4\,H_2SO_4 + 1/2\,C_6H_{12}O_6 \rightarrow Cr_4(OH)_8(SO_4)_2 + 2\,Na_2SO_4 + 3\,CO_2 + 3\,H_2O \text{ Basicity } 66.6\,\%$$
(21)

$$2\,Na_2Cr_2O_7 + 3\,H_2SO_4 + 1/2\,C_6H_{12}O_6 \rightarrow Cr_4(OH)_{10}\,SO_4 + Na_2SO_4 + 3\,CO_2 + H_2O \text{ Basicity } 83.3\,\%$$
(22)

$$2\,Na_2Cr_2O_7 + 2\,H_2SO_4 + 1/2\,C_6H_{12}O_6 \rightarrow Cr_4(OH)_{12} + Na_2SO_4 + 3\,CO_2 \text{ Basicity } 100\,\%$$
(23)

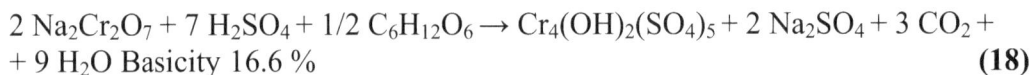

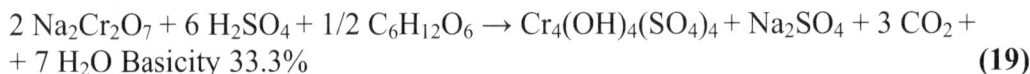

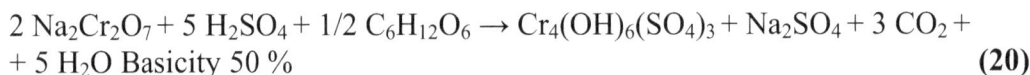

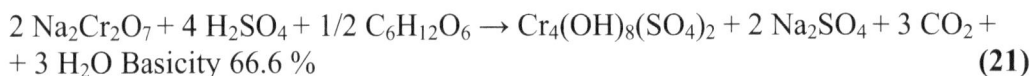

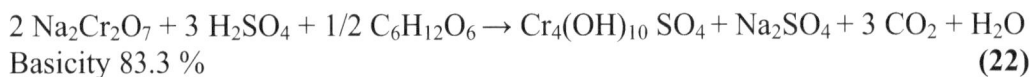

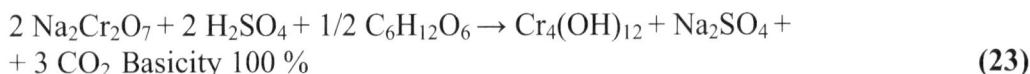

From the above reactions is observed that decreasing the sulfuric acid amount by one mole, basicity of the chromium complex increases at a rate of 16.6%.

Continuing the reaction, we see that when using only two moles of sulfuric acid, required to convert sodium from sodium dichromate into sodium sulfate, basicity of the complex is 100%.

The A amount of sulfuric acid required to obtain, by oxidation-reduction, an amount of basic chromium salt with 100 kg Cr_2O_3 content and basicity b is:

$$A = 258.1 - 193.5\frac{b}{100}$$
(24)

The coefficients of the above equation have resulted from the following calculations: 258.1 is the product of the mole number of sulfuric acid required for reduction of sodium dichromate to sodium sulphate with 0% basicity and obtaining sodium sulfate (4 moles), the number of Cr_2O_3 moles contained in 100 kg Cr_2O_3 (0.6579) and molecular weight of sulfuric acid, as follows:

$$258.1 = 4 \times 0.6579 \times 98.064 = \text{constant}$$
(25)

193.5 is the number of sulfuric acid moles required for reduction of sodium dichromate and obtaining sodium sulfate with 0% basicity (3 moles), the number of Cr_2O_3 moles contained in 100 kg of Cr_2O_3 (0.6579) and molecular weight of sulfuric acid, as follows:

$$193.5 = 3 \times 0.6479 \times 98.064 = \text{constant} \tag{26}$$

If salts of other tanning metals are introduced in the redox system in order to convert hexavalent chromium into trivalent chromium, a series of changes will take place.

Thus, to obtain the chromium-aluminum heterocomplex, aluminum sulphate is used, and the reaction is as follows:

Basic Chromium-Aluminium Heterocomplexes

To obtain a chromium-aluminum heterocomplex with 0% basicity (a double chromium and aluminum sulphate) 8 moles of H_2SO_4 are required, similarly to the sodium sulphate obtained from sodium dichromate. Reducing the amount of sulfuric acid by one mole at a time, a growth rate of basicity of 8.33% can be seen, so that, by eliminating all of the H_2SO_4, only a 66.6% basicity is obtained.

This behavior is due to the fact that during the formation reaction of chromium-aluminum heterocomplexes, the aluminum sulphate releases sulfuric acid, decreasing basicity of heterocomplexes compared to the basicity of chromium complexes synthesized using the same amount of sulfuric acid. The occurrence of the redox reaction in the absence of sulfuric acid is evidence that aluminum sulfate releases sulfuric acid during the reaction. This explains the fact that a higher basicity than 66.6% cannot be obtained for synthesis of chromium-aluminum heterocomplex.

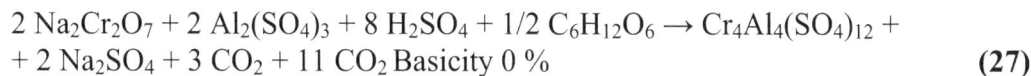

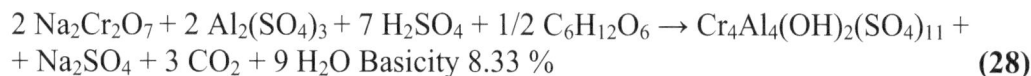

$$2\,Na_2Cr_2O_7 + 2\,Al_2(SO_4)_3 + 8\,H_2SO_4 + 1/2\,C_6H_{12}O_6 \rightarrow Cr_4Al_4(SO_4)_{12} + 2\,Na_2SO_4 + 3\,CO_2 + 11\,CO_2 \text{ Basicity } 0\,\% \tag{27}$$

$$2\,Na_2Cr_2O_7 + 2\,Al_2(SO_4)_3 + 7\,H_2SO_4 + 1/2\,C_6H_{12}O_6 \rightarrow Cr_4Al_4(OH)_2(SO_4)_{11} + Na_2SO_4 + 3\,CO_2 + 9\,H_2O \text{ Basicity } 8.33\,\% \tag{28}$$

$2\ Na_2Cr_2O_7 + 2\ Al_2(SO_4)_3 + 6\ H_2SO_4 + 1/2\ C_6H_{12}O_6 \rightarrow Cr_4Al_4(OH)_4(SO_4)_{10} +$
$+\ 2\ Na_2SO_4 + 3\ CO_2 + 7\ H_2O$ Basicity 16.6 %   (29)

$2\ Na_2Cr_2O_7 + 2\ Al_2(SO_4)_3 + 5\ H_2SO_4 + 1/2\ C_6H_{12}O_6 \rightarrow Cr_4Al_4(OH)_6(SO_4)_9 +$
$+\ 2\ Na_2SO_4 + 3\ CO_2 + 5\ H_2O$ Basicity 25 %   (30)

$2\ Na_2Cr_2O_7 + 2\ Al_2(SO_4)_3 + 4\ H_2SO_4 + 1/2\ C_6H_{12}O_6 \rightarrow Cr_4Al_4(OH)_8(SO_4)_8 +$
$+\ 2\ Na_2SO_4 + 3\ CO_2 + 3\ H_2O$ Basicity 33.3%   (31)

$2\ Na_2Cr_2O_7 + 2\ Al_2(SO_4)_3 + 3\ H_2SO_4 + 1/2\ C_6H_{12}O_6 \rightarrow Cr_4Al_4(OH)_{10}(SO_4)_7 +$
$+\ 2\ Na_2SO_4 + 3\ CO_2 + H_2O$ Basicity 41.66 %   (32)

$2\ Na_2Cr_2O_7 + 2\ Al_2(SO_4)_3 + 2\ H_2SO_4 + 1/2\ C_6H_{12}O_6 \rightarrow Cr_4Al_4(OH)_{12}(SO_4)_6 +$
$+\ 2\ Na_2SO_4 + 3\ CO_2$ Basicity 50 %   (33)

$2\ Na_2Cr_2O_7 + 2\ Al_2(SO_4)_3 + H_2SO_4 + 1/2\ C_6H_{12}O_6 + 3\ H_2O \rightarrow Cr_4Al_4(OH)_{14}(SO_4)_5 +$
$+\ 2\ Na_2SO_4 + 3\ CO_2$ Basicity 58.33 %   (34)

$2\ Na_2Cr_2O_7 + 2\ Al_2(SO_4)_3 + 1/2\ C_6H_{12}O_6 + 5\ H_2O \rightarrow Cr_4Al_4(OH)_{16}(SO_4)_4 +$
$+\ 2\ Na_2SO_4 + 3\ CO_2$ Basicity 66.6 %   (35)

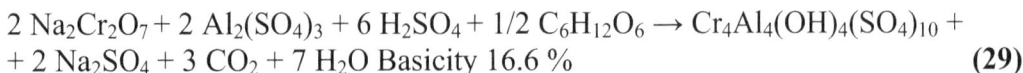

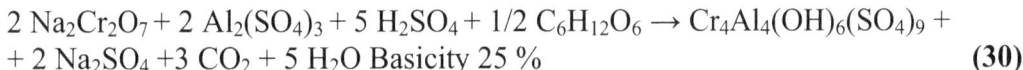

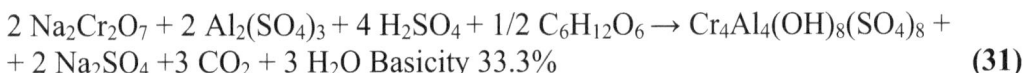

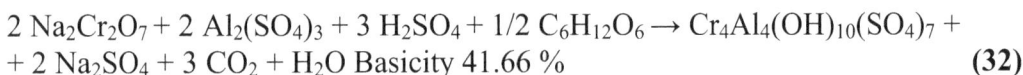

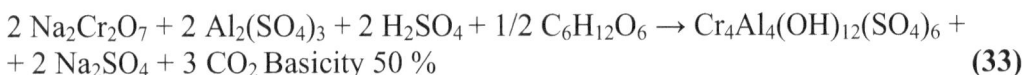

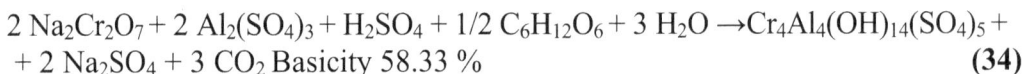

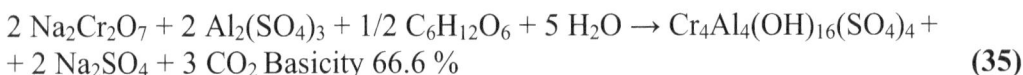

The amount of sulfuric acid (A) needed to obtain a quantity of chromium-aluminum heterocomplex containing 100 kg metal oxides ($Cr_2O_3 + Al_2O_3$) and desired basicity b, can be calculated with the formula:

$$A = 258.1\frac{CCr_2O_3}{100} - \left[193.5\frac{CCr_2O_4}{100} + 288.6\frac{CAl_2O_3}{100}\right]\frac{b}{100} \tag{36}$$

where:

$C\ Cr_2O_3$ is the Cr_2O_3 concentration expressed in %

$C\ Al_2O_3$ is the Al_2O_3 concentration expressed in %.

Coefficients 258.1 and 193.5 are obtained using the same calculations as in the formula of sulfuric acid required to obtain basic chromium sulphate.

Coefficient 288.6 results from multiplying the number of sulfuric acid moles required to obtain aluminum sulphate with 0% basicity, the number of Al_2O_3 moles in 100 kg of Al_2O_3 and molecular weight of sulfuric acid:

$$288.6 = 3 \times 0.981 \times 98.064 = \text{constant} \tag{37}$$

For the synthesis of chromium-iron heterocomplex, ferrous sulphate is introduced in the oxidation-reduction system of hexavalent chromium to trivalent chromium, and the reaction is as follows:

Basic Chromium-Iron Heterocomplexes

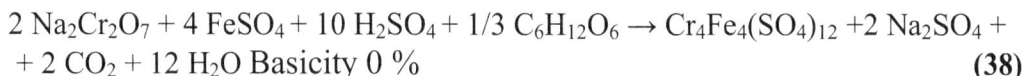

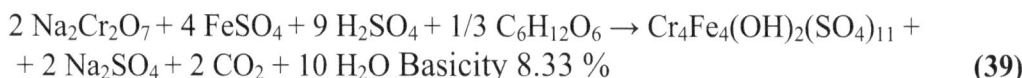

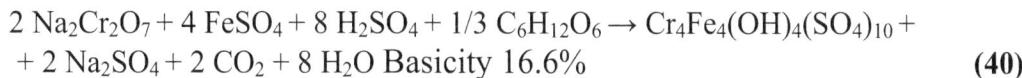

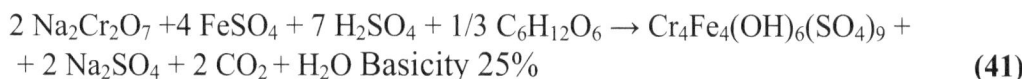

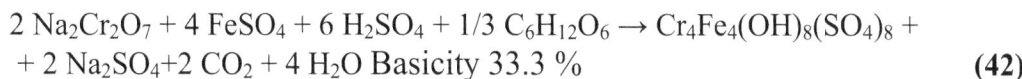

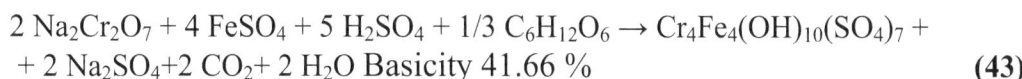

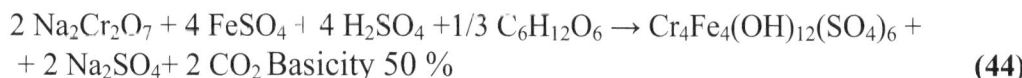

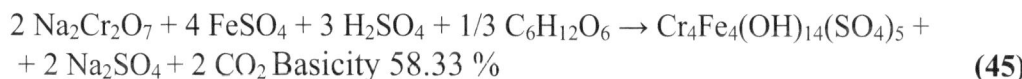

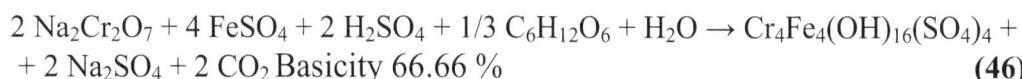

$$2\,Na_2Cr_2O_7 + 4\,FeSO_4 + 10\,H_2SO_4 + 1/3\,C_6H_{12}O_6 \rightarrow Cr_4Fe_4(SO_4)_{12} + 2\,Na_2SO_4 +$$
$$+ 2\,CO_2 + 12\,H_2O \quad \text{Basicity 0 \%} \tag{38}$$

$$2\,Na_2Cr_2O_7 + 4\,FeSO_4 + 9\,H_2SO_4 + 1/3\,C_6H_{12}O_6 \rightarrow Cr_4Fe_4(OH)_2(SO_4)_{11} +$$
$$+ 2\,Na_2SO_4 + 2\,CO_2 + 10\,H_2O \quad \text{Basicity 8.33 \%} \tag{39}$$

$$2\,Na_2Cr_2O_7 + 4\,FeSO_4 + 8\,H_2SO_4 + 1/3\,C_6H_{12}O_6 \rightarrow Cr_4Fe_4(OH)_4(SO_4)_{10} +$$
$$+ 2\,Na_2SO_4 + 2\,CO_2 + 8\,H_2O \quad \text{Basicity 16.6\%} \tag{40}$$

$$2\,Na_2Cr_2O_7 + 4\,FeSO_4 + 7\,H_2SO_4 + 1/3\,C_6H_{12}O_6 \rightarrow Cr_4Fe_4(OH)_6(SO_4)_9 +$$
$$+ 2\,Na_2SO_4 + 2\,CO_2 + H_2O \quad \text{Basicity 25\%} \tag{41}$$

$$2\,Na_2Cr_2O_7 + 4\,FeSO_4 + 6\,H_2SO_4 + 1/3\,C_6H_{12}O_6 \rightarrow Cr_4Fe_4(OH)_8(SO_4)_8 +$$
$$+ 2\,Na_2SO_4 + 2\,CO_2 + 4\,H_2O \quad \text{Basicity 33.3 \%} \tag{42}$$

$$2\,Na_2Cr_2O_7 + 4\,FeSO_4 + 5\,H_2SO_4 + 1/3\,C_6H_{12}O_6 \rightarrow Cr_4Fe_4(OH)_{10}(SO_4)_7 +$$
$$+ 2\,Na_2SO_4 + 2\,CO_2 + 2\,H_2O \quad \text{Basicity 41.66 \%} \tag{43}$$

$$2\,Na_2Cr_2O_7 + 4\,FeSO_4 + 4\,H_2SO_4 + 1/3\,C_6H_{12}O_6 \rightarrow Cr_4Fe_4(OH)_{12}(SO_4)_6 +$$
$$+ 2\,Na_2SO_4 + 2\,CO_2 \quad \text{Basicity 50 \%} \tag{44}$$

$$2\,Na_2Cr_2O_7 + 4\,FeSO_4 + 3\,H_2SO_4 + 1/3\,C_6H_{12}O_6 \rightarrow Cr_4Fe_4(OH)_{14}(SO_4)_5 +$$
$$+ 2\,Na_2SO_4 + 2\,CO_2 \quad \text{Basicity 58.33 \%} \tag{45}$$

$$2\,Na_2Cr_2O_7 + 4\,FeSO_4 + 2\,H_2SO_4 + 1/3\,C_6H_{12}O_6 + H_2O \rightarrow Cr_4Fe_4(OH)_{16}(SO_4)_4 +$$
$$+ 2\,Na_2SO_4 + 2\,CO_2 \quad \text{Basicity 66.66 \%} \tag{46}$$

The formation reactions of chromium-iron heterocomplexes show that to obtain heterocomplexes with 0% basicity, 10 moles of H_2SO_4 are required, therefore an extra 2 moles than in the formation of chromium complexes and chromium-aluminum heterocomplexes.

The addition of 2 moles of sulfuric acid is required for converting ferrous sulphate to ferric sulphate, according to the reaction:

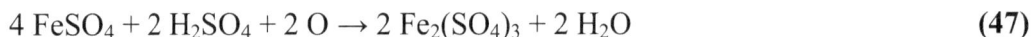

$$4\ FeSO_4 + 2\ H_2SO_4 + 2\ O \rightarrow 2\ Fe_2(SO_4)_3 + 2\ H_2O \tag{47}$$

In this reaction of converting ferrous sulphate to ferric sulphate two oxygen atoms are also consumed, released by sodium dichromate during the reaction of Cr^{6+} reduction to Cr^{3+}, which has the effect of reducing the amount of glucose, which was 1/2 moles for chromium complexes and chromium-aluminum heterocomplexes, to 1/3 moles. So, in this case, the iron in the first part of the reaction acts as a reductant in competition with glucose, and then as a donor of sulfuric acid, which contributes to the reduction of chromium-iron heterocomplex basicity compared to the basicity of chromium complexes obtained using the same amount of sulfuric acid.

The formation reaction of chromium-iron heterocomplexes will occur as in the chromium-aluminum heterocomplexes, except that for the same basicity, the amount of sulfuric acid is 2 moles higher.

The amount of sulfuric acid required to obtain chromium-iron heterocomplexes containing 100 kg of metal oxides ($Cr_2O_3 + Fe_2O_3$) of the desired basicity, b can be calculated using the following formula:

$$A = \left(258.1 \frac{CCr_2O_3}{100} + 61.4 \frac{CFe_2O_3}{100} \right) - \left(193.5 \frac{CCr_2O_3}{100} + 184.20 \frac{CFe_2O_3}{100} \right) \frac{b}{100} \tag{48}$$

where:

CCr_2O_3 is the Cr_2O_3 concentration expressed in %

CFe_2O_3 is the Fe_2O_3 concentration expressed in %.

Coefficients 258.1 and 193.5 are obtained similarly to chromium complexes and chromium-aluminum heterocomplexes.

Coefficient 61.4 is obtained by multiplying the number of sulfuric acid moles required to obtain iron sulfate with 0% basicity (in the composition of chromium-iron heterocomplex), with the number of moles of Fe_2O_3 contained in 100 kg Fe_2O_3 and with the molecular weight of sulfuric acid, as follows:

$$61.4 = 1 \times 0.6261 \times 98.064 = \text{constant.} \tag{49}$$

Coefficient 184.20 is obtained by multiplying the number of sulfuric acid moles required to obtain ferric sulfate with 0% basicity with the number of moles of Fe_2O_3 and contained in 100 kg Fe_2O_3 and with the molecular weight of sulfuric acid, as follows:

$$184.20 = 3 \times 0.6261 \times 98.064 = \text{constant.} \tag{50}$$

In the case where both aluminum sulphate and ferrous sulphate are introduced in the synthesis reaction of chromium complexes of various basicity, the formation mechanism of chromium-aluminum-iron heterocomplexes occurs according to the reactions described below.

Basic Chromium-Aluminium-Iron Heterocomplexes

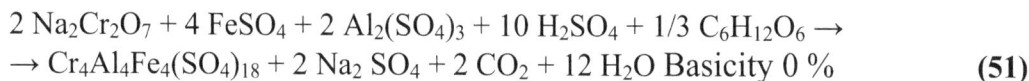

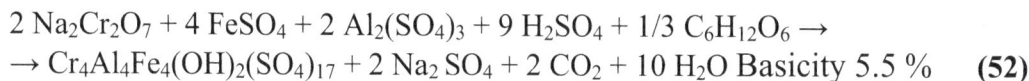

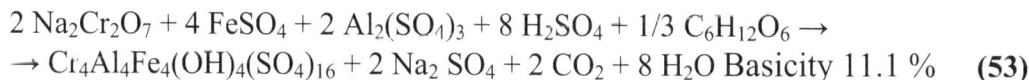

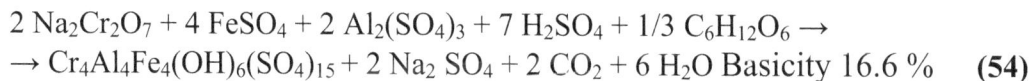

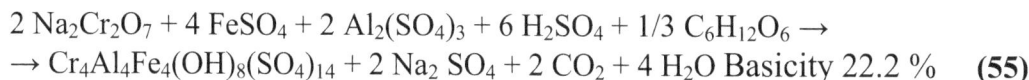

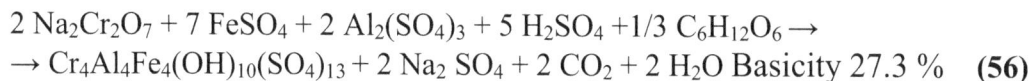

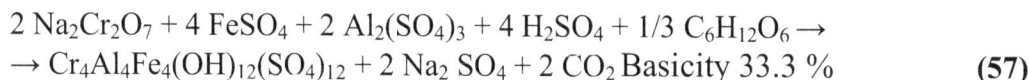

$$2\ Na_2Cr_2O_7 + 4\ FeSO_4 + 2\ Al_2(SO_4)_3 + 10\ H_2SO_4 + 1/3\ C_6H_{12}O_6 \rightarrow$$
$$\rightarrow Cr_4Al_4Fe_4(SO_4)_{18} + 2\ Na_2\ SO_4 + 2\ CO_2 + 12\ H_2O \quad \text{Basicity 0 \%} \tag{51}$$

$$2\ Na_2Cr_2O_7 + 4\ FeSO_4 + 2\ Al_2(SO_4)_3 + 9\ H_2SO_4 + 1/3\ C_6H_{12}O_6 \rightarrow$$
$$\rightarrow Cr_4Al_4Fe_4(OH)_2(SO_4)_{17} + 2\ Na_2\ SO_4 + 2\ CO_2 + 10\ H_2O \quad \text{Basicity 5.5 \%} \tag{52}$$

$$2\ Na_2Cr_2O_7 + 4\ FeSO_4 + 2\ Al_2(SO_4)_3 + 8\ H_2SO_4 + 1/3\ C_6H_{12}O_6 \rightarrow$$
$$\rightarrow Cr_4Al_4Fe_4(OH)_4(SO_4)_{16} + 2\ Na_2\ SO_4 + 2\ CO_2 + 8\ H_2O \quad \text{Basicity 11.1 \%} \tag{53}$$

$$2\ Na_2Cr_2O_7 + 4\ FeSO_4 + 2\ Al_2(SO_4)_3 + 7\ H_2SO_4 + 1/3\ C_6H_{12}O_6 \rightarrow$$
$$\rightarrow Cr_4Al_4Fe_4(OH)_6(SO_4)_{15} + 2\ Na_2\ SO_4 + 2\ CO_2 + 6\ H_2O \quad \text{Basicity 16.6 \%} \tag{54}$$

$$2\ Na_2Cr_2O_7 + 4\ FeSO_4 + 2\ Al_2(SO_4)_3 + 6\ H_2SO_4 + 1/3\ C_6H_{12}O_6 \rightarrow$$
$$\rightarrow Cr_4Al_4Fe_4(OH)_8(SO_4)_{14} + 2\ Na_2\ SO_4 + 2\ CO_2 + 4\ H_2O \quad \text{Basicity 22.2 \%} \tag{55}$$

$$2\ Na_2Cr_2O_7 + 7\ FeSO_4 + 2\ Al_2(SO_4)_3 + 5\ H_2SO_4 + 1/3\ C_6H_{12}O_6 \rightarrow$$
$$\rightarrow Cr_4Al_4Fe_4(OH)_{10}(SO_4)_{13} + 2\ Na_2\ SO_4 + 2\ CO_2 + 2\ H_2O \quad \text{Basicity 27.3 \%} \tag{56}$$

$$2\ Na_2Cr_2O_7 + 4\ FeSO_4 + 2\ Al_2(SO_4)_3 + 4\ H_2SO_4 + 1/3\ C_6H_{12}O_6 \rightarrow$$
$$\rightarrow Cr_4Al_4Fe_4(OH)_{12}(SO_4)_{12} + 2\ Na_2\ SO_4 + 2\ CO_2 \quad \text{Basicity 33.3 \%} \tag{57}$$

$$2\ Na_2Cr_2O_7 + 4\ FeSO_4 + 2\ Al_2(SO_4)_3 + 3\ H_2SO_4 + 1/3\ C_6H_{12}O_6 + 2\ H_2O \rightarrow$$
$$\rightarrow Cr_4Al_4Fe_4(OH)_{14}(SO_4)_{11} + 2\ Na_2\ SO_4 + 2\ CO_2\ \text{Basicity } 38.8\ \% \qquad (58)$$

$$2\ Na_2Cr_2O_7 + 4\ FeSO_4 + 2\ Al_2(SO_4)_3 + 2\ H_2SO_4 + 1/3\ C_6H_{12}O_6 + 4\ H_2O \rightarrow$$
$$\rightarrow Cr_4Al_4Fe_4(OH)_{16}(SO_4)_{10} + 2\ Na_2\ SO_4 + 2\ CO_2\ \text{Basicity } 44.44\ \% \qquad (59)$$

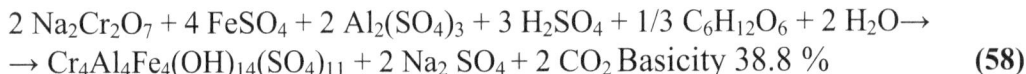

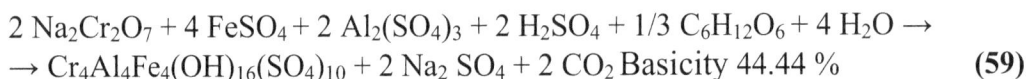

The formation reactions of chromium-aluminum-iron heterocomplexes show that there is a similar behavior to that of chromium-iron heterocomplexes, except that the reduction of a H_2SO_4 mole in the reaction, basicity increase is only 5.55% compared to 8.33%.

This phenomenon is due to the presence of aluminum sulphate, which releases sulfuric acid during the reduction reaction of sodium dichromate. Due to this, maximum basicity which can be reached for chromium-aluminum-iron heterocomplexes is only 44.44%, compared to 66.6%, obtained with chromium-iron heterocomplexes.

The amount of sulfuric acid required to obtain chromium-aluminum-iron heterocomplexes equivalent to 100 kg oxides (Cr_2O_3 + Fe_2O_3 + Al_2O_3) with basicity b, can be calculated using the formula:

$$A = \left(258.1\frac{CCr_2O_3}{100} + 61.4\frac{CFe_2O_3}{100}\right) - \left(193.5\frac{CCr_2O_3}{100} + 288.6\frac{CAl_2O_3}{100} + 184.20\frac{CFe_2O_3}{100}\right)\frac{b}{100} \qquad (60)$$

where:

CCr_2O_3 is the Cr_2O_3 concentration expressed in %

CFe_2O_3 is the Fe_2O_3 concentration expressed in %

CAl_2O_3 is the Al_2O_3 concentration expressed in %.

Numerical coefficients are calculated as in the case of chromium complexes, chromium-aluminum or chromium-iron heterocomplexes.

In the case of chromium reduction in the presence of zirconium sulphate, the formation mechanism of chromium-zirconium heterocomplexes is:

Basic Chromium-Zirconium Heterocomplexes

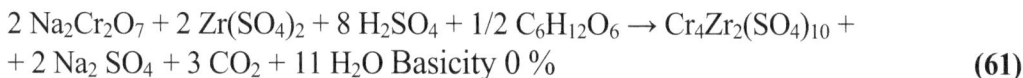

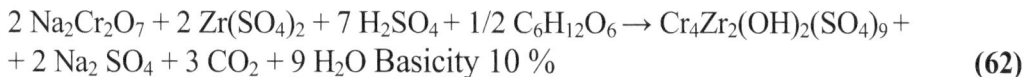

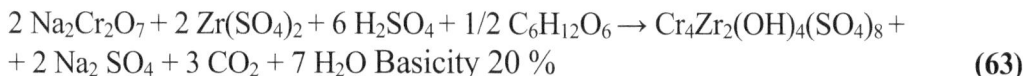

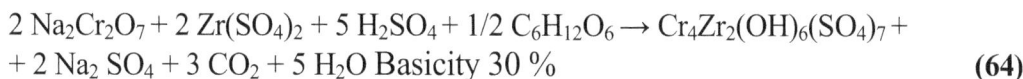

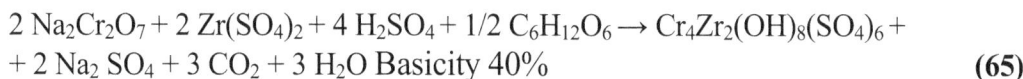

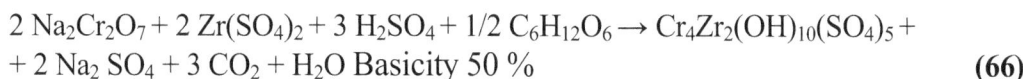

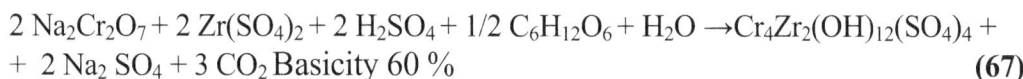

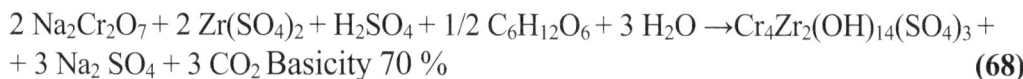

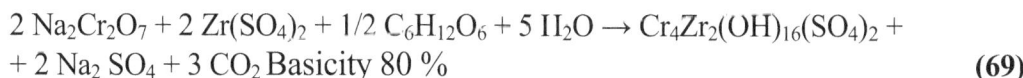

$$2 \, Na_2Cr_2O_7 + 2 \, Zr(SO_4)_2 + 8 \, H_2SO_4 + 1/2 \, C_6H_{12}O_6 \rightarrow Cr_4Zr_2(SO_4)_{10} +$$
$$+ 2 \, Na_2 \, SO_4 + 3 \, CO_2 + 11 \, H_2O \text{ Basicity } 0 \, \% \tag{61}$$

$$2 \, Na_2Cr_2O_7 + 2 \, Zr(SO_4)_2 + 7 \, H_2SO_4 + 1/2 \, C_6H_{12}O_6 \rightarrow Cr_4Zr_2(OH)_2(SO_4)_9 +$$
$$+ 2 \, Na_2 \, SO_4 + 3 \, CO_2 + 9 \, H_2O \text{ Basicity } 10 \, \% \tag{62}$$

$$2 \, Na_2Cr_2O_7 + 2 \, Zr(SO_4)_2 + 6 \, H_2SO_4 + 1/2 \, C_6H_{12}O_6 \rightarrow Cr_4Zr_2(OH)_4(SO_4)_8 +$$
$$+ 2 \, Na_2 \, SO_4 + 3 \, CO_2 + 7 \, H_2O \text{ Basicity } 20 \, \% \tag{63}$$

$$2 \, Na_2Cr_2O_7 + 2 \, Zr(SO_4)_2 + 5 \, H_2SO_4 + 1/2 \, C_6H_{12}O_6 \rightarrow Cr_4Zr_2(OH)_6(SO_4)_7 +$$
$$+ 2 \, Na_2 \, SO_4 + 3 \, CO_2 + 5 \, H_2O \text{ Basicity } 30 \, \% \tag{64}$$

$$2 \, Na_2Cr_2O_7 + 2 \, Zr(SO_4)_2 + 4 \, H_2SO_4 + 1/2 \, C_6H_{12}O_6 \rightarrow Cr_4Zr_2(OH)_8(SO_4)_6 +$$
$$+ 2 \, Na_2 \, SO_4 + 3 \, CO_2 + 3 \, H_2O \text{ Basicity } 40\% \tag{65}$$

$$2 \, Na_2Cr_2O_7 + 2 \, Zr(SO_4)_2 + 3 \, H_2SO_4 + 1/2 \, C_6H_{12}O_6 \rightarrow Cr_4Zr_2(OH)_{10}(SO_4)_5 +$$
$$+ 2 \, Na_2 \, SO_4 + 3 \, CO_2 + H_2O \text{ Basicity } 50 \, \% \tag{66}$$

$$2 \, Na_2Cr_2O_7 + 2 \, Zr(SO_4)_2 + 2 \, H_2SO_4 + 1/2 \, C_6H_{12}O_6 + H_2O \rightarrow Cr_4Zr_2(OH)_{12}(SO_4)_4 +$$
$$+ \; 2 \, Na_2 \, SO_4 + 3 \, CO_2 \text{ Basicity } 60 \, \% \tag{67}$$

$$2 \, Na_2Cr_2O_7 + 2 \, Zr(SO_4)_2 + H_2SO_4 + 1/2 \, C_6H_{12}O_6 + 3 \, H_2O \rightarrow Cr_4Zr_2(OH)_{14}(SO_4)_3 +$$
$$+ 3 \, Na_2 \, SO_4 + 3 \, CO_2 \text{ Basicity } 70 \, \% \tag{68}$$

$$2 \, Na_2Cr_2O_7 + 2 \, Zr(SO_4)_2 + 1/2 \, C_6H_{12}O_6 + 5 \, H_2O \rightarrow Cr_4Zr_2(OH)_{16}(SO_4)_2 +$$
$$+ 2 \, Na_2 \, SO_4 + 3 \, CO_2 \text{ Basicity } 80 \, \% \tag{69}$$

The formation reactions of chromium-zirconium heterocomplexes show that they produce similarly to those of chromium and chromium-aluminum complexes, except that upon the reduction of sulfuric acid amount by 1 mole in reaction, basicity increases by 10% and the final basicity which can be obtained is 80%, which demonstrates that zirconium sulphate releases less sulfuric acid during the reaction, compared with aluminum sulfate.

The amount of sulfuric acid required to obtain chromium-zirconium heterocomplexes equivalent to 100 kg oxides ($Cr_2O_3 + ZrO_2$) with basicity b, can be calculated using the formula:

$$A = \left(258.1\frac{CCr_2O_3}{100}\right) - \left(193.5\frac{CCr_2O_3}{100} + 159.15\frac{CZrO_2}{100}\right)\frac{b}{100} \tag{70}$$

where:

CCr_2O_3 is the Cr_2O_3 concentration expressed in %

$CZrO_2$ is the ZrO_2 concentration expressed in %

Coefficient 159.15 is calculated by multiplying the number of moles of sulfuric acid required to obtain zirconium sulphate with 0% basicity (in the composition of chromium-zirconium heterocomplex), with the number of moles of ZrO_2 in 100 kg of ZrO_2 and with the molecular weight of sulfuric acid:

$$159.15 = 2 \times 0.8115 \times 98.064 = \text{constant.} \tag{71}$$

If aluminum sulfate and zirconium sulfate are introduced in the reduction reaction of sodium dichromate, the formation mechanism of chromium-aluminum-zirconium heterocomplexes is the following:

Basic Chromium-Aluminium-Zirconium Heterocomplexes

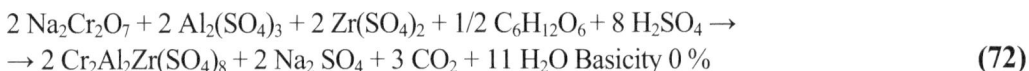

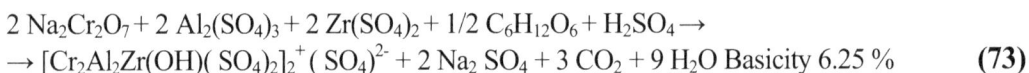

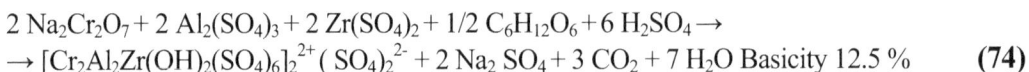

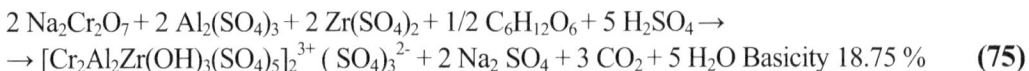

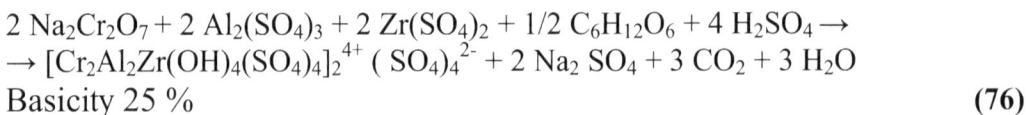

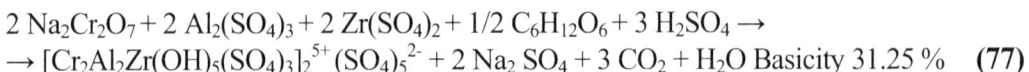

$$2\,Na_2Cr_2O_7 + 2\,Al_2(SO_4)_3 + 2\,Zr(SO_4)_2 + 1/2\,C_6H_{12}O_6 + 8\,H_2SO_4 \rightarrow$$
$$\rightarrow 2\,Cr_2Al_2Zr(SO_4)_8 + 2\,Na_2\,SO_4 + 3\,CO_2 + 11\,H_2O \text{ Basicity } 0\,\% \tag{72}$$

$$2\,Na_2Cr_2O_7 + 2\,Al_2(SO_4)_3 + 2\,Zr(SO_4)_2 + 1/2\,C_6H_{12}O_6 + H_2SO_4 \rightarrow$$
$$\rightarrow [Cr_2Al_2Zr(OH)(SO_4)_2]_2^+ (SO_4)^{2-} + 2\,Na_2\,SO_4 + 3\,CO_2 + 9\,H_2O \text{ Basicity } 6.25\,\% \tag{73}$$

$$2\,Na_2Cr_2O_7 + 2\,Al_2(SO_4)_3 + 2\,Zr(SO_4)_2 + 1/2\,C_6H_{12}O_6 + 6\,H_2SO_4 \rightarrow$$
$$\rightarrow [Cr_2Al_2Zr(OH)_2(SO_4)_6]_2^{2+} (SO_4)_2^{2-} + 2\,Na_2\,SO_4 + 3\,CO_2 + 7\,H_2O \text{ Basicity } 12.5\,\% \tag{74}$$

$$2\,Na_2Cr_2O_7 + 2\,Al_2(SO_4)_3 + 2\,Zr(SO_4)_2 + 1/2\,C_6H_{12}O_6 + 5\,H_2SO_4 \rightarrow$$
$$\rightarrow [Cr_2Al_2Zr(OH)_3(SO_4)_5]_2^{3+} (SO_4)_3^{2-} + 2\,Na_2\,SO_4 + 3\,CO_2 + 5\,H_2O \text{ Basicity } 18.75\,\% \tag{75}$$

$$2\,Na_2Cr_2O_7 + 2\,Al_2(SO_4)_3 + 2\,Zr(SO_4)_2 + 1/2\,C_6H_{12}O_6 + 4\,H_2SO_4 \rightarrow$$
$$\rightarrow [Cr_2Al_2Zr(OH)_4(SO_4)_4]_2^{4+} (SO_4)_4^{2-} + 2\,Na_2\,SO_4 + 3\,CO_2 + 3\,H_2O$$
Basicity 25 % $\tag{76}$

$$2\,Na_2Cr_2O_7 + 2\,Al_2(SO_4)_3 + 2\,Zr(SO_4)_2 + 1/2\,C_6H_{12}O_6 + 3\,H_2SO_4 \rightarrow$$
$$\rightarrow [Cr_2Al_2Zr(OH)_5(SO_4)_3]_2^{5+} (SO_4)_5^{2-} + 2\,Na_2\,SO_4 + 3\,CO_2 + H_2O \text{ Basicity } 31.25\,\% \tag{77}$$

$$2 \ Na_2Cr_2O_7 + 2 \ Al_2(SO_4)_3 + 2 \ Zr(SO_4)_2 + 1/2 \ C_6H_{12}O_6 + 2 \ H_2SO_4 + H_2O \rightarrow$$
$$\rightarrow [Cr_2Al_2Zr(OH)_6(SO_4)_2]_2^{6+} \ (SO_4)_6^{2-} + 2 \ Na_2 \ SO_4 + 3 \ CO_2 \ \text{Basicity 37.5 \%} \quad \textbf{(78)}$$

$$2 \ Na_2Cr_2O_7 + 2 \ Al_2(SO_4)_3 + 2 \ Zr(SO_4)_2 + 1/2 \ C_6H_{12}O_6 + H_2SO_4 + 3 \ H_2O \rightarrow$$
$$\rightarrow [Cr_2Al_2Zr(OH)_7(SO_4)]_2^{7+} \ (SO_4)_7^{2-} + 2 \ Na_2 \ SO_4 + 3 \ CO_2 \ \text{Basicity 43.75 \%} \quad \textbf{(79)}$$

$$Na_2Cr_2O_7 + 2 \ Al_2(SO_4)_3 + 2 \ Zr(SO_4)_2 + 1/2 \ C_6H_{12}O_6 + 5 \ H_2O \rightarrow$$
$$\rightarrow [Cr_2Al_2Zr(OH)_8]_2^{8+} \ (SO_4)_8^{2-} + 2 \ Na_2 \ SO_4 + 3 \ CO_2 \ \text{Basicity 50 \%} \quad \textbf{(80)}$$

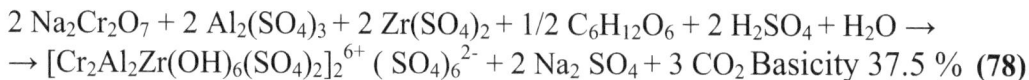

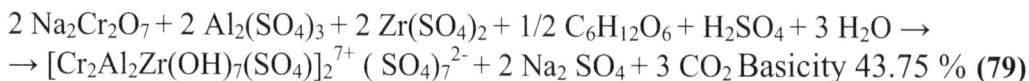

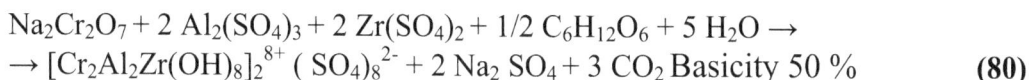

For chromium-aluminum-zirconium heterocomplexes, the synthesis reaction occurs similarly to that of chromium and chromium-aluminum-zirconium ones, except that by reducing the amount of sulfuric acid by 1 mole, basicity increases by 6.25% compared to 8.33% in chromium-aluminum and 10% in chromium-zirconium. This is because both reactants introduced (aluminum sulfate and zirconium sulfate) release sulfuric acid during the reaction. For this reason the maximum basicity which can be achieved is only 50%.

The amount of sulfuric acid required to obtain an amount of chromium-aluminum-zirconium heterocomplex containing 100 kg of metal oxides ($Cr_2O_3 + Al_2O_3 + ZrO_2$) and basicity b, can be calculated according to the formula:

$$A = 258.1 \frac{CCr_2O_3}{100} - \left(193.5 \frac{CCr_2O_3}{100} + 288.6 \frac{CAl_2O_3}{100} + 159.15 \frac{CZrO_2}{100} \right) \frac{b}{100} \quad \textbf{(81)}$$

where:

CCr_2O_3 is the Cr_2O_3 concentration expressed in %

$C \ Al_2O_3$ is the Al_2O_3 concentration expressed in %

$CZrO_2$ is the ZrO_2 concentration expressed in %.

Numerical coefficients were obtained according to calculations of sulfuric acid required to obtain chromium and chromium-aluminum-zirconium heterocomplexes.

The reaction mechanism for the synthesis of chromium-iron-zirconium heterocomplexes with different basicity values is as follows:

Basic Chromium-Iron-Zirconium Heterocomplexes

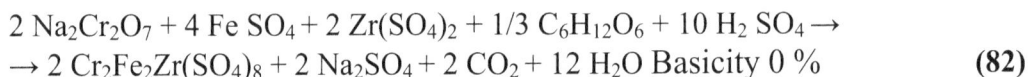

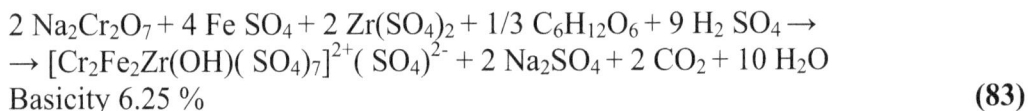

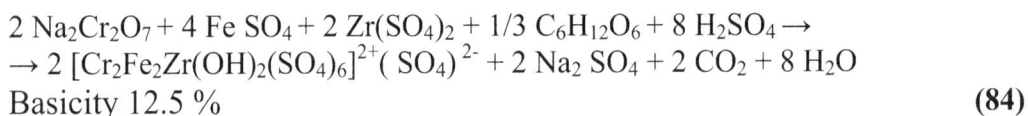

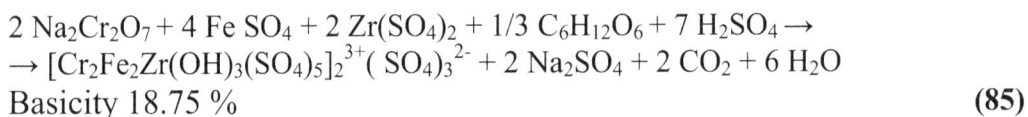

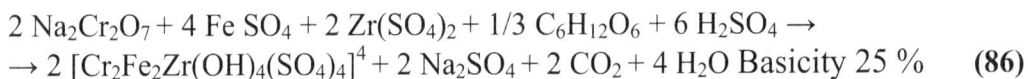

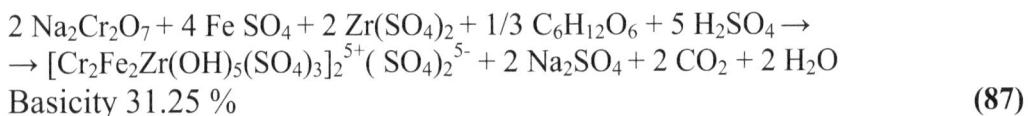

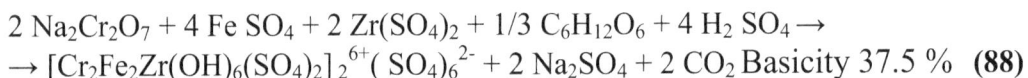

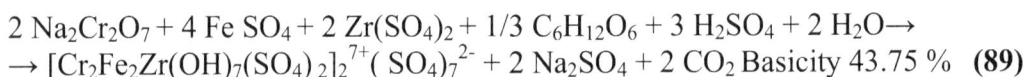

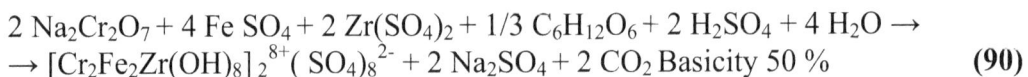

$2\ Na_2Cr_2O_7 + 4\ Fe\ SO_4 + 2\ Zr(SO_4)_2 + 1/3\ C_6H_{12}O_6 + 10\ H_2\ SO_4 \rightarrow$
$\rightarrow 2\ Cr_2Fe_2Zr(SO_4)_8 + 2\ Na_2SO_4 + 2\ CO_2 + 12\ H_2O$ Basicity 0 % $\qquad$ (82)

$2\ Na_2Cr_2O_7 + 4\ Fe\ SO_4 + 2\ Zr(SO_4)_2 + 1/3\ C_6H_{12}O_6 + 9\ H_2\ SO_4 \rightarrow$
$\rightarrow [Cr_2Fe_2Zr(OH)(SO_4)_7]^{2+}(SO_4)^{2-} + 2\ Na_2SO_4 + 2\ CO_2 + 10\ H_2O$
Basicity 6.25 % $\qquad$ (83)

$2\ Na_2Cr_2O_7 + 4\ Fe\ SO_4 + 2\ Zr(SO_4)_2 + 1/3\ C_6H_{12}O_6 + 8\ H_2SO_4 \rightarrow$
$\rightarrow 2\ [Cr_2Fe_2Zr(OH)_2(SO_4)_6]^{2+}(SO_4)^{2-} + 2\ Na_2\ SO_4 + 2\ CO_2 + 8\ H_2O$
Basicity 12.5 % $\qquad$ (84)

$2\ Na_2Cr_2O_7 + 4\ Fe\ SO_4 + 2\ Zr(SO_4)_2 + 1/3\ C_6H_{12}O_6 + 7\ H_2SO_4 \rightarrow$
$\rightarrow [Cr_2Fe_2Zr(OH)_3(SO_4)_5]_2^{3+}(SO_4)_3^{2-} + 2\ Na_2SO_4 + 2\ CO_2 + 6\ H_2O$
Basicity 18.75 % $\qquad$ (85)

$2\ Na_2Cr_2O_7 + 4\ Fe\ SO_4 + 2\ Zr(SO_4)_2 + 1/3\ C_6H_{12}O_6 + 6\ H_2SO_4 \rightarrow$
$\rightarrow 2\ [Cr_2Fe_2Zr(OH)_4(SO_4)_4]^4 + 2\ Na_2SO_4 + 2\ CO_2 + 4\ H_2O$ Basicity 25 % $\quad$ (86)

$2\ Na_2Cr_2O_7 + 4\ Fe\ SO_4 + 2\ Zr(SO_4)_2 + 1/3\ C_6H_{12}O_6 + 5\ H_2SO_4 \rightarrow$
$\rightarrow [Cr_2Fe_2Zr(OH)_5(SO_4)_3]_2^{5+}(SO_4)_2^{5-} + 2\ Na_2SO_4 + 2\ CO_2 + 2\ H_2O$
Basicity 31.25 % $\qquad$ (87)

$2\ Na_2Cr_2O_7 + 4\ Fe\ SO_4 + 2\ Zr(SO_4)_2 + 1/3\ C_6H_{12}O_6 + 4\ H_2\ SO_4 \rightarrow$
$\rightarrow [Cr_2Fe_2Zr(OH)_6(SO_4)_2]_2^{6+}(SO_4)_6^{2-} + 2\ Na_2SO_4 + 2\ CO_2$ Basicity 37.5 % $\quad$ (88)

$2\ Na_2Cr_2O_7 + 4\ Fe\ SO_4 + 2\ Zr(SO_4)_2 + 1/3\ C_6H_{12}O_6 + 3\ H_2SO_4 + 2\ H_2O \rightarrow$
$\rightarrow [Cr_2Fe_2Zr(OH)_7(SO_4)_2]_2^{7+}(SO_4)_7^{2-} + 2\ Na_2SO_4 + 2\ CO_2$ Basicity 43.75 % $\quad$ (89)

$2\ Na_2Cr_2O_7 + 4\ Fe\ SO_4 + 2\ Zr(SO_4)_2 + 1/3\ C_6H_{12}O_6 + 2\ H_2SO_4 + 4\ H_2O \rightarrow$
$\rightarrow [Cr_2Fe_2Zr(OH)_8]_2^{8+}(SO_4)_8^{2-} + 2\ Na_2SO_4 + 2\ CO_2$ Basicity 50 % $\qquad$ (90)

The formation reactions of chromium-iron-zirconium heterocomplexes with different basicity values show similar behavior to synthesis of chromium-aluminum-zirconium heterocomplexes, except that in all cases an additional 2 moles of sulfuric acid are used, which, as already shown, are required for transformation of ferrous sulphate into ferric sulphate.

The amount of sulfuric acid required to obtain an amount of chromium-iron-zirconium heterocomplex containing 100 kg of metal oxides ($Cr_2O_3 + Fe_2O_3 + ZrO_2$) and basicity b, can be calculated according to the formula:

$$A = \left(258.1\frac{CCr_2O_3}{100} + 61.4\frac{CFe_2O_3}{100}\right) - \left(193.5\frac{CCr_2O_3}{100} + 184.19\frac{CFe_2O_3}{100} + 159.15\frac{ZrO_2}{100}\right)\frac{b}{100} \quad \textbf{(91)}$$

where:

CCr_2O_3 is the Cr_2O_3 concentration expressed in %

$C Fe_2O_3$ is the Fe_2O_3 concentration expressed in %

$CZrO_2$ is the ZrO_2 concentration expressed in %,

and coefficients are calculated as explained in the case of chromium or chromium-zirconium-iron heterocomplexes.

Comparing synthesis mechanisms for several types of tanning metallic heterocomplexes allowed the identification of how each reaction system can influence basicity of the developed tanning agent. These influences are summarized in Table **6**.

Table 6: Variation of basicity of tanning metallic heterocomplexes compared to chromium complexes and depending on the amount of sulfuric acid

Amount of sulfuric acid, moles	Basicity, %						
	Complexes	Heterocomplexes:					
	Cr	Cr-Al	Cr-Fe	Cr-Al-Fe	Cr-Zr	Cr-Al-Zr	Cr-Fe-Zr
10 H$_2$SO$_4$	-	-	0	0	-	-	0
9 H$_2$SO$_4$	-	-	8.33	5.55	-	-	6.25
8 H$_2$SO$_4$	0	0	16.66	11.11	0	0	12.50
7 H$_2$SO$_4$	16.66	8.33	25.00	16.66	10.00	6.25	16.75
6 H$_2$SO$_4$	33.33	16.66	33.00	22.22	20.00	12.50	25.50
5 H$_2$SO$_4$	50.00	25.00	41.66	27.77	30.00	18.75	31.25
4 H$_2$SO$_4$	66.66	33.33	50.00	33.33	40.00	25.00	37.50
3 H$_2$SO$_4$	83.20	41.66	58.33	38.88	50.00	31.25	43.75
2 H$_2$SO$_4$	100.00	50.00	66.66	44.44	60.00	37.50	50.00
1 H$_2$SO$_4$	-	58.33	-	-	70.00	43.75	-
0 H$_2$SO$_4$	-	66.66	-	-	80.00	50.00	-

Table **6** shows that, in all cases, 8 moles of sulfuric acid are required for the formation of tanning metallic heterocomplexes with 0% basicity, except those containing iron, which require 10 moles of sulfuric acid.

By reducing the amount of sulfuric acid with 1 mol, the highest basicity increase rate is that of the chromium complex, 16.66%, followed by a rate of 10% of chromium-zirconium heterocomplex, 8.33% for chromium-aluminum and chromium-iron heterocomplexes, 6.25% for chromium-iron-zirconium, and 5.55% for chromium-aluminum-iron heterocomplex.

Maximum theoretical basicity values which can be obtained are 100% for chromium, 80% for chromium-aluminum and chromium-iron heterocomplexes, 50% for chromium-aluminum-zirconium and chromium-iron-zirconium heterocomplexes, and 44.4% for chromium-aluminum-iron heterocomplexes.

Based on these data, the results of the tanning metallic heterocomplex synthesis mechanism theoretical study may explain their different acidity values, analytical and electrochemical characteristics which directly influence their structure and their behavior in the leather tanning process.

SYNTHESIS OF HETEROCOMPLEXES BASED ON CHROMIUM, ALUMINIUM, IRON OR ZIRCONIUM

Synthesis of Chromium-Iron Heterocomplexes

The oxidation-reduction reaction for the synthesis of chromium-iron heterocomplexes occurs with Cr^{6+} reduction to Cr^{3+} and oxidation of Fe^{2+} to Fe^{3+} from sodium dichromate and ferrous sulphate in sulfuric acid medium using glucose as main reductant.

Ferrous sulphate is an ecologic, inexpensive product, acting as a reductant for sodium dichromate. Obtaining the chromium-iron heterocomplex with 33% basicity, typical of basic chromium salts, was done according to the reactions listed below, which may result in different types of heterocomplex salts, depending on the possibilities of the sulfate entering the coordination sphere, as it can be seen from the reactions listed below:

$$Na_2Cr_2O_7 + 2\ FeSO_4 + 3\ H_2SO_4 + 1/6\ C_6H_{12}O_6 \rightarrow 2\ [CrFe(OH)_2]^{4+}(SO_4)_2^{2-} +$$
$$+ Na_2SO_4 + CO_2 + 2\ H_2O \tag{92}$$

or

$$\rightarrow 2\ [CrFe(OH)_2(SO_4)]^{2+}SO_4^{2-} + Na_2SO_4 + CO_2 + 2\ H_2O \tag{93}$$

or

$$\rightarrow 2\ [CrFe(OH)_2(SO_4)_2] + Na_2SO_4 + CO_2 + 2H_2O \tag{94}$$

or

$$\rightarrow 2\ [CrFe(OH)_2(SO_4)_3]^{2-}2Na^+ CO_2 + 2\ H_2O \tag{95}$$

From the above reaction it can be noticed that in the synthesis of complex with basicity of 33%, the solution may contain a variety of differently charged heterocomplexes: cationic, anionic or neutral.

For the reaction it is necessary to establish the required amount of sulfuric acid, taking into account that oxidation of $FeSO_4$ into $Fe_2(SO_4)_3$ requires sulfuric acid according to the following reaction:

$$2\ FeSO_4 + H_2SO_4 + O \rightarrow Fe_2(SO_4)_3 + H_2O \tag{96}$$

Therefore, the redox reaction requires sulfuric acid for both reduction of sodium dichromate and for transformation of ferrous sulphate into ferric sulphate.

Theoretically, the reaction should be carried out as follows: sulfuric acid is added to sodium dichromate solution and then the ferrous sulphate solution is added, which will reduce some of the sodium dichromate and will oxidize to ferric sulphate.

Then the reduction reaction is continued using glucose as a reductant until the end of Cr^{6+} reduction to Cr^{3+}.

The required reagents are established based on the stoichiometric calculation of the reaction, as follows:

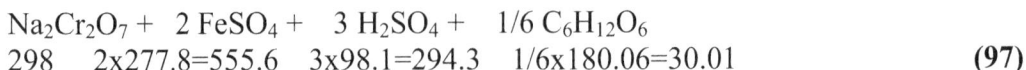

$$Na_2Cr_2O_7 + \quad 2\ FeSO_4 + \quad 3\ H_2SO_4 + \quad 1/6\ C_6H_{12}O_6$$

$$298 \quad 2x277.8{=}555.6 \quad 3x98.1{=}294.3 \quad 1/6x180.06{=}30.01 \tag{97}$$

therefore:

$$152g\ Cr_2O_3\ \text{will correspond to}\ 159\ g\ Fe_2O_3$$

So, for a total of 311.7g metal oxides of which 48.78% Cr_2O_3 and 51.22% Fe_2O_3, according to the synthesis reaction, 294.3g H_2SO_4 100% and 30.2g glucose are needed.

For laboratory scale synthesis of a tanning metallic heterocomplex solution containing 10g metal oxides with different proportions of metal oxides, reagent amounts summarized in Table **7** are used, according to the above indicated stoichiometric calculations.

Synthesis reaction is performed as follows: sulfuric acid is added cold to sodium dichromate solution while stirring and ferrous sulfate is then added. The mixture is heated up to 80°C and the glucose solution is added in thin stream. The completion of oxidation-reduction reaction is verified by using the reaction of potassium iodide in hydrochloric acid medium and in the presence of starch.

Compared to theoretical calculations, synthesis reactions occur with sulfuric acid in excess. The attempt to reduce the amount of sulfuric acid in the synthesis of chromium-iron heterocomplex solution with oxide ratio of 80/20% leads to destabilizing the solution due to increased basicity of complexes.

In conclusion, the experiment shows that in order to achieve the synthesis of chromium-iron heterocomplex, an excess of sulfuric acid is needed.

The stability of chromium-iron metallic heterocomplex solutions is determined by measuring the pH at initial and final precipitation points by treatment with NaOH 0.5n, NaHCO$_3$ 0.5n and Na$_2$CO$_3$ 0.2n solutions. The results are shown in Table **6**. Table **7** shows a very good stability of chromium-iron heterocomplex solutions, much improved compared to that of a basic iron solution, which is around a pH of 2-2.2.

Table 7: Reagent amounts used in the synthesis of chromium-iron tanning metallic heterocomplex solutions.

Cr-Fe heterocomplex solution	Amount of reagents used					
$Cr_2O_3/Fe_2O_3\%$	Cr_2O_3 (g)	Sodium dichromate solution 119.5 g/l Cr_2O_3 (ml)	Fe_2O_3 (g)	Ferrous sulfate (g)	H_2SO_4 95% (g)	$C_6H_{12}O_6$ (g)
80/20	8	66.9	2	7	13	5.6
70/30	7	58.6	3	10.5	13	4.9
60/40	7	50.2	4	14	14	4.2
50/50	5	41.8	5	17.5	16	3.5
40/60	4	33.45	6	21	19	2.8

The conclusion that can be drawn from the data in Table **8** is that heterocomplexation improves stability properties of iron as a result of the action of alkaline substances.

Table 8: pH of chromium-iron heterocomplex synthesized solutions and their behaviour in alkaline medium.

Solution name	pH of solution as such	pH of analytical solution	pH at precipitation points with:					
			NaOH 0.5n		NaHCO$_3$ 0.5n		Na$_2$CO$_3$ 0.2n	
			initial	final	initial	final	initial	final
Cr-Fe 80/20%	2.2	3.1	5.8	6.0	5.7	6.3	5.8	6.2
Cr-Fe 70/30%	1.8	2.5	5.5	5.9	5.9	6.0	5.8	6.0
Cr-Fe 60/40%	1.5	2.0	5.5	5.8	6.0	6.0	5.8	5.9
Cr-Fe 50/50%	0.8	1.6	5.5	5.7	5.9	5.9	5.7	5.8
Cr-Fe 40/60%	0.3	1.6	4.6	5.9	5.6	5.7	5.5	5.8

Table **8** also indicates that, as Cr_2O_3 content decreases in the heterocomplex, the Fe_2O_3 content increases, while pH of the solution as such and pH of the analytical solution decrease. In contrast, pH values at precipitation points by alkalinization remain high in all cases, indicating good resistance of these products to alkaline treatment and therefore, to an increase in basicity.

This is particularly important because it allows raising the pH of these solutions to 3.6-4.2, the necessary values for manufacturing technology in tanneries, without the risk of precipitation.

To check the behavior of heterocomplex solutions to aging, a chromium-iron complex was synthesized with metal oxides ratio of 80/20%; the pH values and points of precipitation were determined for its solution after preparation and after storage for 5 months in ambient conditions. The first finding is that after 5 months of storage, the solution is perfectly clear, with no sediment.

Table **9** presents the pH values of the solution and pH at precipitation points initially, and after 5 months of storage.

From Table **9** it is observed that after 5 months of storage, the pH of the solution as such and that of the analytical solution undergo a slight increase, from 1.8 to 2.2 and from 3.1 to 3.4, respectively. Normally we should expect a decrease in pH during storage due to a tendency of hydrolysis over time of complex salts of chromium and especially those of iron.

Table 9: pH of chromium-iron heterocomplex solutions with metal oxide ratio of 80/20 %, initially, after 5 months of storage and at precipitation points.

Moment of measurement	pH solution	pH solution	pH at precipitation point:			
			$NaHCO_3$ 0.5n		Na_2CO_3 0.2n	
	as such	analytical	initial	final	initial	final
initial (24 hours after synthesis)	1.8	3.1	5.7	6.0	6.0	6.0
after 5 months of storage	2.2	3.4	5.9	5.9	5.6	5.8

In this case, probably, instead of hydrolysis followed by release of acid, complexing phenomena occurred, acid groups entering the complex, resulting in an increase of pH.

What is important is that heterocomplex stability remained generally constant and high upon the action of alkalis.

Given their complexity, chemical characterization of the synthesized solutions was performed using several methods. These methods were marked as follows:

- M – iron masking and iodometric determination of Cr_2O_3;

- F – precipitation of iron and iodometric determination of Cr_2O_3;

- D – determination of iron by the difference between total oxides and Cr_2O_3;

- C – calcination of precipitate iron at 1000°C and gravimetric determination;

- T – complexometric titration of iron redissolved from the precipitate.

The results of chemical analyses are presented in Table **10**.

Table **10** shows that analytically determined values for Cr_2O_3, Fe_2O_3 and metal oxides, expressed in grams, are close to those used in the synthesis based on theoretical calculations.

Table **10** also shows that the method of determining Cr_2O_3 by masking the iron oxide followed by iodometric titration of chromium is the best method.

The best method for determining Fe_2O_3 is that of determination by difference, involving gravimetric determination of total oxides and Cr_2O_3 by iodometric titration.

Basicity variation of chromium-iron heterocomplex solutions is determined by acid solutions. Thus, a higher percentage of iron oxide in the heterocomplex required a higher acidity, and thus basicity reached negative values.

Table 10: Chemical analysis of chromium-iron heterocomplex solutions.

Type of hetero-complex	Chemical test method used Cr_2O_3	Content of Cr_2O_3 (g/l)	Cr_2O_3 (g)	Chemical test method used Fe_2O_3	Fe_2O_3 (g/l)	Fe_2O_3 (g)	Metal oxides (g/l)	Metal oxides (g)	Basicity (%)
	M	52.19	7.82	D	17.93	2.68	70.12	10.51	
Cr-Fe	F	52.59	7.88	C	17.60	2.60	70.19	10.52	34.4
80/20 %	F	54.61	8.16	T	11.59	1.73	66.20	9.95	
	M	47.02	7.06	D	23.06	3.45	70.08	10.51	
Cr-Fe	F	50.67	7.60	C	23.68	3.55	74.35	11.15	12.89
70/30 %	F	46.61	6.99	T	20.28	3.04	66.89	10.03	
	M	38.91	5.83	D	29.41	4.41	68.32	10.24	
Cr-Fe	F	34.45	5.36	C	26.08	3.91	60.53	9.07	8.20
60/40 %	F	36.48	5.47	T	26.08	3.91	62.56	9.38	
	M	31.82	4.77	D	37.06	5.55	68.88	10.33	
Cr-Fe	F	32.43	4.86	C	31.60	4.74	64.03	9.60	-25.2
50/50 %	F	34.45	5.16	T	31.88	4.78	66.33	9.94	
	M	26.04	3.90	D	41.32	6.19	67.36	10.10	
Cr-Fe	F	26.35	3,95	C	41.04	6.15	67.39	10.10	-46.43
40/60 %	F	26.35	3.95	T	40.47	6.08	66.92	10.03	

These basicity values are brought to technological values acceptable for the tanning process by buffering solutions with solutions of alkali until they reach the specific pH for tanning solutions (pH 2.8-3.2).

In conclusion, using the oxidation-reduction method to convert Cr^{6+} to Cr^{3+} and Fe^{2+} to Fe^{3+} leads to chromium-iron heterocomplexes with different ratios between Cr_2O_3 and Fe_2O_3, and basicity values influenced by the amount of acid used in the reaction.

Determining precipitation points, the pH values at the initial and final precipitation points, shows a very good stability of chromium-iron heterocomplex to alkali action and very good stability over time, therefore indicating the possibility of their use in leather tanning and retanning.

Synthesis of Chromium-Aluminum-Iron Heterocomplexes

Chromium-aluminum-iron tanning heterocomplexes is also synthesized by oxidation-reduction reaction for the transformation of Cr^{6+} to Cr^{3+} and Fe^{2+} to

Fe^{3+}. This reaction occurs in the presence of both aluminum sulphate and ferrous sulphate.

The synthesis of this product is more complicated because it must be taken into account that during the reaction, ferrous sulphate is a consumer of sulfuric acid, just like chromium. Also, during the reaction, aluminum sulfate is a donor of sulfuric acid, just like ferric sulfate formed by oxidation-reduction. In this case establishing the quantity of sulfuric acid is much more difficult.

Given these factors, it was found that the chromium-aluminum-iron heterocomplex formation reaction could take place as follows:

$$Na_2Cr_2O_7 + Al_2(SO_4)_3 + 2\ FeSO_4 + 2\ H_2SO_4 + 1/6\ C_6H_{12}O_6 \rightarrow$$
$$\rightarrow 2[CrAlFe(OH)_3]^{6+}(SO_4)_3{}^{2-} + Na_2SO_4 + CO_2 \tag{98}$$

or

$$\rightarrow 2[CrAlFe(OH)_3(SO_4)]^{4+}(SO_4)_2{}^{2-} + Na_2SO_4 + CO_2 \tag{99}$$

or

$$\rightarrow 2[CrAlFe(OH)_3(SO_4)_2]^{2+}SO_4{}^{2-} + Na_2SO_4 + CO_2 \tag{100}$$

or

$$\rightarrow 2[CrAlFe(OH)_3(SO_4)_3]^0 + Na_2SO_4 + CO_2 \tag{101}$$

or

$$\rightarrow 2[CrAlFe(OH)_3(SO_4)_4]^{-2}.2Na^+ + CO_2 \tag{102}$$

The reactions listed above show that chromium-aluminum-iron heterocomplexes which are cationic, anionic and without charge can form in the solution, with 33.33% basicity.

Chromium-aluminum-iron heterocomplexes can be synthesized by adding sodium dichromate solution to aluminum sulfate solution while stirring, then concentrated sulfuric acid, ferrous sulphate and glucose solution are added when reaching the temperature of 85°C.

The necessary amount of reagents is established based on stoichiometric calculations, according to the synthesis reaction, as follows:

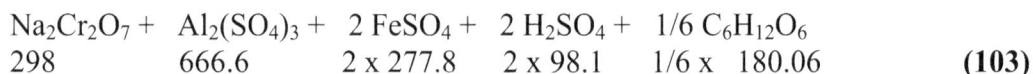

$$Na_2Cr_2O_7 + Al_2(SO_4)_3 + 2\ FeSO_4 + 2\ H_2SO_4 + 1/6\ C_6H_{12}O_6$$

298   666.6   2 x 277.8   2 x 98.1   1/6 x 180.06   **(103)**

So, according to the reaction we have a total of 413.7g metal oxides (Cr_2O_3, Al_2O_3, Fe_2O_3) representing 36.8% Cr_2O_3, 24.6% Al_2O_3 and 38.6% Fe_2O_3.

According to the reaction, for 413.7g metal oxides 196.2g sulfuric acid 100% and 30.01 g glucose are necessary.

Synthesis of chromium-aluminum-iron heterocomplexes with different ratios of Cr_2O_3, Al_2O_3 and Fe_2O_3 was chosen as follows:

-   chromium-aluminum-iron heterocomplexes containing 50% Cr_2O_3, 35% Al_2O_3 and 15% Fe_2O_3;

-   chromium-aluminum-iron heterocomplexes containing 40% Cr_2O_3, 40% Al_2O_3 and 20% Fe_2O_3;

-   chromium-aluminum-iron heterocomplexes containing 33.3% Cr_2O_3, 33.3% Al_2O_3 and 33.3% Fe_2O_3;

-   chromium-aluminum-iron heterocomplexes containing 30% Cr_2O_3, 45% Al_2O_3 and 25% Fe_2O_3.

The quantities of reagents used for the four types of syntheses are presented in Table **11**. In the case of chromium-aluminum-iron heterocomplex with oxide ratios of 50/35/15%, 4 variants were synthesized in which the amount of sulfuric acid was reduced from 15 grams to 9 grams.

Syntheses are performed as follows: the sodium dichromate solution, the amount of sulfuric acid and finally ferrous sulfate are added cold to the aluminum sulfate solution. The mixture is heated to 85°C and then the calculated amount of glucose is added in thin stream, raising the temperature to boiling point. After 30-40 minutes of boiling, the completion of reduction is checked using potassium iodide, starch and hydrochloric acid.

Table 11: Reagent amounts used in chromium-aluminum-iron heterocomplex synthesis

Chromium-aluminum-iron heterocomplex solution $Cr_2O_3/Al_2O_3/Fe_2O_3\%$	Amount of reagents							
	Cr_2O_3 (g)	Sodium dichromate solution with 119.5g/l Cr_2O_3 (ml)	Al_2O_3 (g)	Aluminum sulfate solution with 60.8g/l Al_2O_3 (ml)	Fe_2O_3 (g)	Ferrous solution (g)	Sulfuric acid 95 % (g)	Glucose (g)
50/35/15	5	41.8	3.5	57.7	1.5	5.3	15 12 10 9	3.5
40/40/20	4	33.4	4	65.8	2	7	10	2.8
33.3/33.3/33.3	3.3	27.8	3.3	54.8	3.3	11.6	13.0	2.3
30/45/25	3	25.1	4.5	74.0	2.5	8.7	13.0	2

Table **12** presents pH values of synthesized chromium-aluminum-iron heterocomplex solutions and pH values of the solutions at precipitation points after NaOH 0.5n, NaHCO$_3$ 0.5n and Na$_2$CO$_3$ 0.2n treatment.

Table 12: Chromium-aluminum-iron heterocomplexes, pH of solutions and pH at precipitation points

Solution name	Amount of H_2SO_4 94% used in synthesis (g)	pH of solution as such	pH of analytical solution	pH at precipitation point with:					
				NaOH 0.5n		NaHCO$_3$ 0.5n		Na$_2$CO$_3$ 0.2n	
				initial	final	initial	final	initial	final
Cr-Al-Fe 50/35/15 %	15	1.0	1.8	4.3	4.6	4.3	4.8	4.5	4.6
Cr-Al-Fe 50/35/15 %	12	1.5	2.2	4.0	4.6	4.3	4.3	4.2	5.2
Cr-Al-Fe 50/35/15 %	10	1.6	2.3	4.2	4.6	4.5	4.5	4,0	4.6
Cr-Al-Fe 50/35/15 %	9	2.0	3.1	3.8	4.9	4.3	5.2	4.2	4.6
Cr-Al-Fe 40/40/20 %	10	1.5	2.1	4.0	4.3	4.3	4.3	3.9	4.3
Cr-Al-Fe 33/33/33%	13	0.5	1.7	3.9	4.6	4.3	5.8	4.0	4.5
Cr-Al-Fe 30/45/25 %	13	0.3	1.7	4.0	4.4	4.3	5.9	3.8	5.5

Table **12** shows that for the 4 synthesized chromium-aluminum-iron heterocomplex solutions with ratio of metal oxides of 50/35/15%, as the amount of sulfuric acid reduced, the pH of the solutions as such progressively increased from 1 to 2, and the pH of analytical solutions increased from 1.8 to 3.1.

Also, acidity increases with decreasing amount of Cr_2O_3 in the chromium-aluminum-iron heterocomplex solution composition and their stability to the interaction with alkali is good, which indicates the possibility of their buffering and use in the specific conditions of the tanning process.

Regarding the pH of solutions ar precipitation points, with one insignificant exception, they all have values above 4, which means that they can be used in tanning without any problems. It may be noted, however, that the pH of the chromium-aluminum-iron heterocomplex solution at points of precipitation is noticeably lower than for the chromium-iron heterocomplexes. This indicates a lower stability of the former compounds. It follows, therefore, that the presence of aluminum in the heterocomplex structure leads to a decrease in its stability.

The decrease in stability of chromium-aluminum-iron heterocomplex may be primarily due to the higher affinity of chromium to iron. The role of "ligand" of aluminum, known in the literature in the case of chromium-aluminum combinations can manifest in this case as well, actually destabilizing the bond between chromium and iron.

To observe its behavior over time, a chromium-aluminum-iron heterocomplex solution with metal oxide ratio of 50/35/15% was stored for 5 months under normal storage conditions. The pH values of solutions as such, analytical solutions and pH values at precipitation points were determined for the initial solution and for the solution aged for 5 months, and the results are presented in Table **13**.

As with the chromium-iron heterocomplexes, during storage of chromium-aluminum-iron heterocomplex solutions a slight increase in the pH of the solution as such and the analytical solution is found, indicating, in this case as well, that during aging there is an increasing degree of complexation of the heterocomplex.

Table 13: Chromium-aluminum-iron heterocomplexes; pH of solutions and pH of solutions at precipitation points initially and after 5 months of storage.

Solution	pH as such	pH of analytical solution	pH at precipitation point with:			
			NaOH 0.5n		Na₂CO₃ 0.2n	
			initial	final	initial	final
Cr-Al-Fe 50/35/15% initial	0.8	2.0	3.8	6.2	5.0	5.0
Cr-Al-Fe 50/35/15% after 5 months	1.2	2.2	4.0	5.8	4.0	4.8

With respect to pH values at precipitation points, no significant changes are noted during aging, indicating that stability of chromium-aluminum-iron heterocomplex is preserved over time.

Organoleptically, it was found that these solutions remained perfectly clear without presenting any sediment.

Chemical analyses of chromium-aluminum-iron heterocomplex solutions are summarized in Table **14**.

To determine the metal oxides, the same methods of analysis as in the case of chromium-iron heterocomplex were used, marked as follows:

- M – iron masking and iodometric determination of Cr_2O_3;

- F – precipitation of iron with sodium peroxide, precipitate filtration and iodometric determination of Cr_2O_3;

- D – gravimetric determination of total oxides, of which Al_2O_3 is determined by difference;

- C – calcination of precipitate iron at 1000°C and gravimetric determination;

- T – complexometric titration of iron redissolved and separated from the precipitate.

Table **14** shows that the four chromium-aluminum-iron heterocomplex solutions with metal oxides ratio of 50/35/15%, the analytically determined quantities of Cr_2O_3, Al_2O_3, Fe_2O_3 and total metal oxides are close to those calculated theoretically for synthesis.

Table 14: Chemical analysis of chromium-aluminum-iron heterocomplex solutions.

Solution	Cr_2O_3 (g/l)	Cr_2O_3 (g)	Al_2O_3 (g/l)	Al_2O_3 (g)	Fe_2O_3 (g/l)	Fe_2O_3 (g)	Metal oxides (g/l)	Metal oxides (g)	Basicity (%)
Cr-Al-Fe 50/35/15%	M 31.76	4.76	D 29.04	4.35	C 11.60	1.74	C 72.40	10.86	
	F 36.41	5.46	D 21.50	3.22	T 14.49	2.17	-	-	-25.97
	F 34.38	5.15	-	-	-	-	-	-	
Cr-Al-Fe 50/35/15%	M 31.55	4.73	D 25.77	3.86	C 10.56	1.58	C 67.88	10.18	
	F 36.41	5.46	D 19.88	2.98	T 11.59	1.73	-	-	0.88
	F 34.38	5.15	-	-	-	-	-	-	
Cr-Al-Fe 50/35/15%	M 31.15	4.67	D 30.25	4.53	C 11.12	1.66	C 72.5	10.87	
	F 32.36	4.85	D 26.02	3.90	T 14.49	2.17	-	-	6.76
	F 36.41	5.46	-	-	-	-	-	-	
Cr-Al-Fe 50/35/15%	M 31.55	4.73	D 22.85	3.42	C 14.40	2.16	C 68.80	10.32	10.62
	F 34.38	5.15	D 19.05	2.85	T 17.39	2.60	-	-	
	F 32.35	4.85	-	-	-	-	-	-	
Cr-Al-Fe 40/40/20%	M 26.09	3.91	D 26.09	3.91	C 15.28	2.29	C 67.76	10.16	23.0
	F 26.29	3.94	D 26.98	4.04	T 14.49	2.17	-	-	
	F 26.29	3.94	-	-	-	-	-	-	
Cr-Al-Fe 33/33/33%	M 21.89	3.28	D 24.83	3.72	C 22.80	3.42	C 69.52	10.42	
	F 22.29	3.34	D 24.05	3.60	T 23.18	3.47	-	-	-23.7
	F 22.29	3.31	-	-	-	-	-	-	
Cr-Al-Fe 30/45/25%	M 19.24	2.88	D 34.29	5.14	C 16.18	2.52	C 69.12	10.36	
	F 20.22	3.03	D 28.58	4.28	T 20.28	3.08	-	-	-39.41
	F 18.20	2.76	-	-	-	-	-	-	

Regarding basicity, Table **14** shows that when a large amount of sulfuric acid is used in synthesis, due to high acidity of the solution, basicity is negative. As the amount of sulfuric acid used is reduced, a gradual increase in basicity can be seen.

In the case of these heterocomplexes, the low pH of the initial solution is not an impediment for its use in the tanning process because, given the good stability to alkali treatment, the pH can be raised to the optimum values for use in tanning without the risk of changing stability.

In conclusion, using the oxidation-reduction method to convert Cr^{6+} to Cr^{3+} and Fe^{2+} to Fe^{3+} in a reaction medium with sulfuric acid and aluminum sulfate, heterocomplexes containing chromium-aluminum-iron with different ratios of metal oxides can be synthesized. Also, various methods of analysis are available for determining basicity and metal oxide content of developed heterocomplexes.

Determination of pH at initial and final precipitation points indicates a stability of these heterocomplexes which recommends them for use in leather tanning and retanning.

Synthesis of Chromium-Iron-Zirconium Heterocomplexes

Sodium dichromate, ferrous sulphate and zirconium sulfate can be used for chromium-iron-zirconium heterocomplex synthesis.

Two reductants coexist in the oxidation-reduction reaction: glucose and ferrous sulfate. The reaction requires additional consumption of sulfuric acid due to oxidation of ferrous sulphate to ferric sulphate:

$$2\ FeSO_4 + H_2SO_4 + O\ \rightarrow\ Fe_2(SO_4)_3 + H_2O \tag{104}$$

Also, there are two sulfuric acid generators: zirconium sulfate and ferric sulfate formed during the reaction.

In theory, the chromium-iron-zirconium heterocomplex reaction formation occurs according to reactions given below:

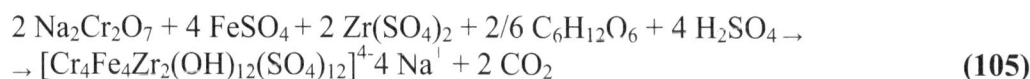

$$2\ Na_2Cr_2O_7 + 4\ FeSO_4 + 2\ Zr(SO_4)_2 + 2/6\ C_6H_{12}O_6 + 4\ H_2SO_4 \rightarrow$$
$$\rightarrow [Cr_4Fe_4Zr_2(OH)_{12}(SO_4)_{12}]^{4-} 4\ Na^+ + 2\ CO_2 \tag{105}$$

or

$$\rightarrow [Cr_4Fe_4Zr_2(OH)_{12}(SO_4)_{11}]^{2-} 2\ Na^+ + 2\ CO_2 + Na_2SO_4 \tag{106}$$

or

$$\rightarrow [Cr_4Fe_4Zr_2(OH)_{12}(SO_4)_{10}]^{0} + 2\ CO_2 + 2\ Na_2SO_4 \tag{107}$$

or

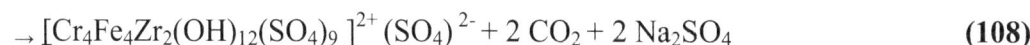

$$\rightarrow [Cr_4Fe_4Zr_2(OH)_{12}(SO_4)_9]^{2+} (SO_4)^{2-} + 2\ CO_2 + 2\ Na_2SO_4 \tag{108}$$

or

$$\rightarrow [Cr_4Fe_4Zr_2(OH)_{12}(SO_4)_8]^{4+} (SO_4)_2{}^{2-} + 2\ CO_2 + 2\ Na_2SO_4 \tag{109}$$

or

$$\rightarrow [Cr_4Fe_4Zr_2(OH)_{12}(SO_4)_7]^{6+} (SO_4)_3{}^{2-} + 2\ CO_2 + 2\ Na_2SO_4 \tag{110}$$

or

$$\rightarrow [Cr_4Fe_4Zr_2(OH)_{12}(SO_4)_6]^{8+} (SO_4)_4{}^{4-} + 2\ CO_2 + 2\ Na_2SO_4 \tag{111}$$

or

$$\rightarrow [Cr_4Fe_4Zr_2(OH)_{12}(SO_4)_5]^{10+} (SO_4)_5{}^{2-} + 2\ CO_2 + 2\ Na_2SO_4 \tag{112}$$

or

$$\rightarrow [Cr_4Fe_4Zr_2(OH)_{12}(SO_4)_4]^{12+} (SO_4)_6{}^{2-} + 2\ CO_2 + 2\ Na_2SO_4 \tag{113}$$

or

$$\rightarrow [Cr_4Fe_4Zr_2(OH)_{12}(SO_4)_3]^{14+} (SO_4)_7{}^{2-} + 2\ CO_2 + 2\ Na_2SO_4 \tag{114}$$

or

$$\rightarrow [Cr_4Fe_4Zr_2(OH)_{12}(SO_4)_2]^{16+} (SO4)_8{}^{2-} + 2\ CO_2 + 2\ Na_2SO_4 \tag{115}$$

or

$$\rightarrow [Cr_4Fe_4Zr_2(OH)_{12}(SO_4)]^{18+} (SO_4)_9{}^{2-} + 2\ CO_2 + 2\ Na_2SO_4 \tag{116}$$

or

$$\rightarrow [Cr_4Fe_4Zr_2(OH)_{12}]^{20+} (SO_4)_{10}{}^{2-} + 2\ CO_2 + 2\ Na_2SO_4 \tag{117}$$

In the case of chromium-iron-zirconium heterocomplexes as well, a variety of compounds with different electrical charges are formed, all showing basicity of 37.5%.

To determine synthesis possibilities and characteristics of chromium-iron-zirconium heterocomplexes, 13 types of hetrocomplexes were obtained with different ratios of metal oxides.

Metal oxide ratios of synthesized chromium-iron-zirconium heterocomplexes were: 80/10/10%, 70/20/10%, 70/10/20%, 60/30/10%, 60/20/20%, 60/10/30%, 50/40/10%, 50/30/20%, 50/20/30%, 50/10/40%, 40/40/20%, 40/30/30% and 40/20/40%.

In all cases 10 g metal oxides were added, with the above indicated proportions. At the end of synthesis, all solutions were brought to the same volume of 100ml.

Given the complexity and variety of these reactions, in order to determine the necessary amount of sulfuric acid to be used in the synthesis, stoichiometric calculations were made from the chromium-iron-zirconium heterocomplex preparation reaction with 31.25% basicity.

Chromium-iron-zirconium heterocomplex preparation reaction with basicity of 31.25% is as follows:

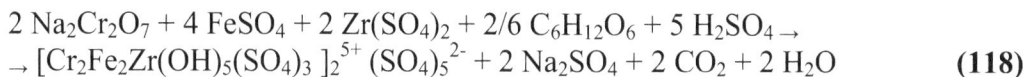

$$2\ Na_2Cr_2O_7 + 4\ FeSO_4 + 2\ Zr(SO_4)_2 + 2/6\ C_6H_{12}O_6 + 5\ H_2SO_4 \rightarrow$$
$$\rightarrow [Cr_2Fe_2Zr(OH)_5(SO_4)_3\ ]_2^{5+}\ (SO_4)_5^{2-} + 2\ Na_2SO_4 + 2\ CO_2 + 2\ H_2O \qquad \textbf{(118)}$$

Corresponding to this synthesis reaction, metal oxide ratios, necessary amounts of sulfuric acid and glucose were calculated as follows:

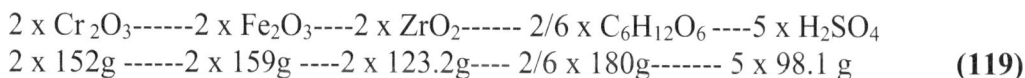

$$2\ x\ Cr_2O_3\text{------}2\ x\ Fe_2O_3\text{----}2\ x\ ZrO_2\text{------}\ 2/6\ x\ C_6H_{12}O_6\text{----}5\ x\ H_2SO_4$$
$$2\ x\ 152g\ \text{------}2\ x\ 159g\ \text{----}2\ x\ 123.2g\text{----}\ 2/6\ x\ 180g\text{-------}\ 5\ x\ 98.1\ g \qquad \textbf{(119)}$$

It follows then, that for the preparation of 10 grams of metal oxides containing 35% Cr_2O_3, 28.4% ZrO_2 and 36.6% Fe_2O_3, the amount of 5.6g H_2SO_4 100% must be added to get a basicity of 31.25%.

It is clear that, by changing the ratio of metal oxides and heterocomplex basicity, the necessary amount of sulfuric acid changes accordingly.

For this reason in the case of chromium-iron-zirconium heterocomplex synthesis, the proportions of each metal oxides were probed to determine the necessary amount of sulfuric acid.

Table **15** presents the quantities of reagents used in chromium-iron-zirconium heterocomplex synthesis and the obtained pH values of solutions.

Table 15: Reagent amounts used in the synthesis of chromium-iron-zirconium heterocomplexes and pH of obtained solutions.

Cr-Fe-Zr heterocomplex solution $Cr_2O_3/Fe_2O_3/ZrO_2$ %	Amount of reagents								pH of solution at 100 ml volume
	Cr_2O_3 (g)	Sodium dichromate solution with 119.5 g/l Cr_2O_3 (ml)	Fe_2O_3 (g)	Ferrous sulfate (g)	ZrO_2 (g)	Zirconium sulfate solution with 103.4 g/l ZrO_2 (ml)	H_2SO_4 95% (g)	$C_6H_{12}O_6$ (g)	
Cr-Fe-Zr 80/10/10 %	8	66.9	1	3.5	1	9.7	10	5.6	2.9
Cr-Fe-Zr 70/20/10 %	7	58.6	2	7	1	9.7	11	4.9	1.7
Cr-Fe-Zr 60/30/10 %	6	50.2	3	10.5	1	9.7	12	4.2	0.7
Cr-Fe-Zr 50/40/10 %	5	41.8	4	14	1	9.7	19	3.5	0.00
Cr-Fe-Zr 70/10/20 %	7	58.6	1	3.5	2	19.3	6	4.9	2.6
Cr-Fe-Zr 60/20/20 %	6	50.2	2	7	2	19.3	10	4.2	0.6
Cr-Fe-Zr 50/30/20 %	5	41.8	3	10.5	2	19.3	12	3.5	0.3
Cr-Fe-Zr 40/40/20 %	4	33.5	4	14	2	19.3	22	2.8	0.0
Cr-Fe-Zr 60/10/30 %	6	50.2	1	3.5	3	29.0	6	4.2	1.1
Cr-Fe-Zr 50/20/30 %	5	41.8	2	7	3	29.0	11	3.5	0.2
Cr-Fe-Zr 40/30/30 %	4	33.5	3	10.5	3	29.0	10	2.8	0.1
Cr-Fe-Zr 50/10/40 %	5	41.8	1	3.5	4	38.7	10	3.5	0.2
Cr-Fe-Zr 40/20/40 %	4	33.5	2	7	4	38.7	7	2.8	0.5

Table **16** presents the results obtained in determining the precipitation points of synthesized chromium-iron-aluminum solutions.

The last items in Table **16** are ferrous sulfate, zirconium sulfate and their mixture in proportions of 50% Fe_2O_3 and 50% ZrO_2.

It is found that ferrous sulfate as such has a very good resistance to alkali, and the pH at precipitation point is between 5.2 and 6.4. In contrast, zirconium sulfate has a poor resistance to alkalis, and pH at precipitation points ranges between 1.2 and 3.3.

The mixture of these two substances in proportion of 50% Fe_2O_3 and 50% ZrO_2 indicates pH values at the point of precipitation ranging between 2.0 and 3.5, which are much lower than those of ferrous sulphate and much closer to those of zirconium sulfate.

Table 16: pH of solutions and pH of solutions at precipitation points for chromium-iron-zirconium heterocomplexes.

Solution	pH solution as such	pH analytical solution	pH at precipitation points with:					
			NaOH 0.5n		NaHCO₃ 0.5n		Na₂CO₃ 0.2r	
			initial	final	initial	final	initial	final
Cr-Fe-Zr 80/10/10%	2.9	3.5	5.5	5.6	6.1	6.1	5.8	5.8
Cr-Fe-Zr 70/20/10%	1.7	2.7	4.7	5.3	6.0	6.0	5.3	5.3
Cr-Fe-Zr 60/30/10%	0.7	2.0	3.1	4.3	5.2	5.2	4.7	4.8
Cr-Fe-Zr 50/40/10%	0.0	1.2	2.3	4.1	4.6	4.6	3.1	4.6
Cr-Fe-Zr 70/10/20%	2.6	3.2	5.1	5.1	5.5	6.1	5.8	5.8
Cr-Fe-Zr 60/20/20%	0.6	1.8	3.5	4.7	5.4	5.4	5.7	4.8
Cr-Fe-Zr 50/30/20%	0.3	1.6	4.2	4.6	5.4	5.4	5.0	5.0
Cr-Fe-Zr 40/40/20%	0.0	1.0	2.3	4.4	2.7	2.7	4.5	4.5
Cr-Fe-Zr 60/10/30%	1.1	2.2	4.0	4.4	6.3	6.3	5.2	5.2
Cr-Fe-Zr 50/20/30%	0.2	1.5	2.5	3.7	4.8	4.8	4.5	4.5
Cr-Fe-Zr 40/30/30%	0.1	1.1	2.4	3.8	4.4	4.4	4.3	4.3
Cr-Fe-Zr 50/10/40%	0.2	1.2	2.5	3.6	4.3	4.3	3.2	4.0
Cr-Fe-Zr 40/20/40%	0.5	1.6	3.0	3.7	4.5	4.5	3.8	4.3
Ferrous sulfate	2.8	4.2	5.2	5.7	6.4	6.4	6.3	6.3
Zirconium sulfate	0.0	1.1	1.2	1.8	2.8	2.8	3.3	3.3
Ferrous sulfate + Zirconium sulfate with 50%Fe₂O₃+50%ZrO₂	0.2	1.3	2.0	3.1	3.1	3.1	2.4	3.5

The Cr^{6+} to Cr^{3+} redox reaction also leads to the complexation process of the 3 tanning metals forming tanning metallic heterocomplexes with superior stability values compared to homeopolynuclear complex salts. With the increase of iron proportion in heterocomplexes, it is noticed that pH values at precipitation points decrease and alkali resistance decreases, while remaining good.

From the point of view of stability over time of chromium-iron-zirconium heterocomplexes, Table **17** shows a considerable decrease of pH at precipitation points after 20 days storage, indicating a reduction in resistance to alkali.

Even in the case of the solution stored for 20 days, chromium-iron-zirconium metal heterocomplexed are stable to alkalis, which recommends them for use in leather tanning.

Following synthesis of chromium-iron-zirconium heterocomplexes and their stability analysis it is found that a wide range of products suitable for tanning and retanning can be obtained.

Table 17: pH of solutions and pH of solutions at precipitation points after 20 days of storage for chromium-iron-zirconium heterocomplexes.

| Solution | pH solution as such | pH analytical solution | pH at precipitation points with: | | | | | |
| | | | NaOH 0.5n | | NaHCO$_3$ 0.5n | | Na$_2$CO$_3$ 0.2n | |
			initial	final	initial	final	initial	final
Cr-Fe-Zr 80/10/10 %	2.7	3.2	4.8	5.2	5.6	5.6	5.4	5.4
Cr-Fe-Zr 70/20/10 %	1.7	2.6	4.5	5.0	5.1	5.1	4.8	4.8
Cr-Fe-Zr 60/30/10 %	0.8	2.1	4.0	4.5	4.8	4.8	4.6	4.6
Cr-Fe-Zr 50/40/10 %	0.0	1.2	3.9	4.4	4.4	4.4	4.2	4.3
Cr-Fe-Zr 70/10/20 %	2.5	3.0	4.8	5.3	5.7	5.7	5.1	5.4
Cr-Fe-Zr 60/20/20 %	0.5	1.8	3.5	4.1	4.2	4.2	2.9	4.1
Cr-Fe-Zr 50/30/20 %	0.2	1.4	2.9	4.5	4.2	4.2	4.2	4.2
Cr-Fe-Zr 40/40/20 %	0.0	1.0	2.7	3.7	2.4	2.4	4.0	4.0
Cr-Fe-Zr 60/10/30 %	1.3	2.3	3.8	4.6	5.1	5.1	4.8	4.8
Cr-Fe-Zr 50/20/30 %	0.3	1.4	3.0	3.5	4.1	4.1	3.9	4.08
Cr-Fe-Zr 40/30/30 %	0.1	1.2	2.1	3.5	4.1	4.1	4.0	4.0
Cr-Fe-Zr 50/10/40 %	0.1	1.3	2.5	3.6	3.7	3.7	2.5	3.8
Cr-Fe-Zr 40/20/40 %	0.5	1.6	3.5	3.8	4.1	4.1	4.1	4.1

Synthesis of tanning heterocomplexes based on chromium, iron, aluminum and zirconium can be achieved by oxidation-reduction reaction in sulfuric acid medium in the presence of glucose as a reducing agent. Stability to alkali of tanning metallic heterocomplexes is superior to the stability of component metal salts and can be ranked as follows:

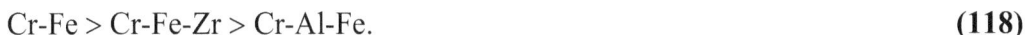

$$Cr\text{-}Fe > Cr\text{-}Fe\text{-}Zr > Cr\text{-}Al\text{-}Fe. \tag{118}$$

In trimetallic heterocomplexes, the increase of aluminum (Cr-Al-Fe) or iron (Fe-Cr-Zr) proportions leads to a decrease in stability to alkali, which remains within the values used in leather tanning and retanning. The stability of tanning metallic heterocomplex solutions varies very little over time (20 days-5 months), without significantly affecting stability to alkalis. Laboratory scale synthesis of tanning metallic heterocomplexes confirmed the elaborated theoretical hypotheses of stoichiometric mechanisms of reaction.

Tanning metallic heterocomplexs allow reduction of chromium oxide offer in tanning by 30-50% and a synergism of interaction with collagen generated by synthesis.

Pilot Scale Synthesis of Chromium-Iron Tanning Metallic Heterocomplexes

The product containing 70% Cr_2O_3 and 30% Fe_2O_3 was selected in order to establish and check the pilot scale production technology for chromium-iron heterocomplexes. We chose this heterocomplex taking into account the pH of solutions obtained, the good resistance to alkalis and reduced chromium consumption which can be achieved in tanneries.

In order to establish the synthesis technology it was necessary to determine the amount of sulfuric acid that is used in the reaction, also taking into account the fact that for $FeSO_4$ to turn into $Fe_2(SO_4)_3$ sulfuric acid is required as follows:

1 mol of H_2SO_4 for 2 moles of $FeSO_4$.

So, in the redox reaction sulfuric acid is needed both for the reduction of sodium dichromate, and for transformation of ferrous sulphate into ferric sulphate.

Theoretically, the reaction should be conducted as follows: sulfuric acid is added to the sodium dichromate solution, then the ferrous sulphate solution is added, which will act as a reductant for the sodium dichromate and will turn it into a ferric salt. Then the reduction reaction is continued using glucose as reductant, until complete conversion of Cr^{6+} to Cr^{3+}.

This method, successfully used during laboratory scale research, was tested at pilot scale in order to obtain a solution containing 2200g metal oxides, of which 1540g Cr_2O_3 and 660g Fe_2O_3. To achieve the synthesis we used the quantities of reagents summarized in Table **18**.

Table 18: Amounts of reagents used in pilot scale synthesis of chromium-iron heterocomplexes.

Chromium-iron heterocomplex solution Cr_2O_3/Fe_2O_3 (%)	Amounts of reagents used					
	Cr_2O_3 (g)	Sodium dichromate solution with 216.3g/l Cr_2O_3 (ml)	Fe_2O_3 (g)	Ferrous sulfate (g)	H_2SO_4 95% (g)	$C_6H_{12}O_6$ (g)
70/30	1540	7100	660	2310	2800	1078

The method used for synthesis is:

-   the sodium dichromate solution is prepared by dissolving it in 200% hot water and then Cr_2O_3 content is determined analytically;

-   ferrous sulfate is dissolved in 200% hot water, in which 420 g H_2SO_4 were introduced (representing 15% of the required amount of H_2SO_4 for reduction);

-   glucose is dissolved in 200% water at 70-80°C.

2380 g of sulfuric acid in 94% concentration is added in thin stream to sodium dichromate solution heated to 80°C, then, while continuously stirring, the previously obtained system consisting of ferrous sulphate and sulfuric acid is added. The reaction is continued by adding the warm glucose solution in thin stream.

Synthesis takes place with abundant heat and gas release.

After completion of effervescent phase, the solution is further heated to 95-100°C for 30-45 minutes.

The end of the reduction reaction is checked organoleptically, from the color of the final solution and by iodometric testing, namely by treatment with potassium iodide, starch and hydrochloric acid. The final pH of the solution thus obtained is 1.8.

The next day, the solution is buffered by continuous stirring with 8 liters of sodium bicarbonate solution of 5% concentration, dosed in thin stream up to a pH value of 2.5. The amount of sodium bicarbonate used is about 18% of the amount of metal oxides.

The chemical characteristics of the chromium-iron heterocomplex solution are listed in Table **19**.

Table 19: Chemical characteristics for the chromium-iron heterocomplex solution.

Chemical characteristics	MU	Value
Total metal oxides	g/l	116.1
Cr_2O_3	g/l	35.5
Fe_2O_3	g/l	29.5
Basicity	%	29.5
pH of solution as such		2.55
pH of analytical solution		3.10

Determination of precipitation points and of pH at precipitation points led to the results shown in Table **20**.

The analysis of data presented in Table **20** shows that solutions of chromium-iron heterocomplexes have good resistance to alkalis and a pH value which allows their use in tanneries.

Given that metal heterocomplex solutions undergo transformation over time due to hydrolysis, olification and polymerization phenomena, they are made into powder by atomization. Making chromium-iron heterocomplexes in the form of powder also allows performing structural analyses such as: IR spectroscopy, RES, SEM, EDAX *etc.*

For the solution to undergo drying by atomization, it is prepared by filtering using ordinary filter paper and heating to 40-45°C. Drying parameters are: inlet temperature: 250-300°C, outlet temperature: 120-125°C, air pressure: 4 atmospheres.

Chromium-iron heterocomplex powders thus obtained are chemically analyzed and the results are given in Table **21**.

Table 20: Precipitation points and pH values at precipitation points for the chromium-iron heterocomplex solution.

Precipitation points at treatment with NaOH 0.5n	
initial	0.8ml pH=3.59
final	3.8ml pH=4.90
Precipitation points at treatment with NaHCO₃ 0.5n	
initial	1.2ml pH=5.02
final	2.9ml pH=5.24
Precipitation points at treatment with Na₂CO₃ 0.2n	
initial	0.9ml pH=4.90
final	2ml pH=5.10

Table 21: Chemical characteristics for chromium-iron heterocomplex powders.

Characteristics	MU	Value
Moisture	%	4.03
Metal oxides	%	28.9*
Cr_2O_3	%	18.4*
Fe_2O_3	%	10.5*
Basicity	%	32.14
pH of solution 1:10		2.5
pH of analytical solution		2.6

*recalculated for moisture-free

Results of determining precipitation points and pH of precipitation points of the chromium-iron heterocomplex analytical solution are given in Table **22**.

Comparing the results obtained at precipitation points of the initial chromium-iron heterocomplex solution and of the solution obtained by dissolving the heterocomplex powder, we observe two important differences, namely: pH values of the solutions obtained from powders are higher than the initial ones, indicating a greater resistance to alkali; for these solutions, initial and final precipitation points and their corresponding pH values are identical, while for the initial solution, in all cases the values are different. This behavior leads to the conclusion that during drying by atomization, sulfate groups are forced to enter the complex and masked complexes are developed with high resistance to alkali action.

Table 22: Precipitation points for the chromium-iron heterocomplex solution obtained from powder.

Precipitation points with NaOH 0.5n: initial final	14ml pH=5.10 14ml pH=5.10
Precipitation points with NaHCO$_3$ 0.5n: initial final	8.6ml pH=5.6 8.6ml pH=5.6
Precipitation points with Na$_2$CO$_3$ 0.2n: initial final	7.5ml pH=5.53 7.5ml pH=5.53

Pilot Scale Synthesis of Chromium-Aluminum-Iron Tanning Metallic Heterocomplexes

In order to check the pilot technology, the solution containing 50% Cr_2O_3, 35% Al_2O_3 and 15% Fe_2O_3 is chosen, the most convenient in terms of stability to alkalis and possibilities of chromium oxide reduction.

For pilot testing, a solution of chromium-aluminum-iron heterocomplexes containing 2200g metal oxides, of which 1100g Cr_2O_3, 770g Al_2O_3 and 330g Fe_2O_3 is synthesized.

To achieve the synthesis, the quantities of reagents summarized in Table **23** are used.

The work method is the following:

- the sodium dichromate solution is prepared by dissolving it in 200% hot water and Cr_2O_3 content is determined analytically;

- the aluminum sulfate solution is prepared by dissolving it in 200% hot water and Al_2O_3 content is determined analytically;

- ferrous sulfate is dissolved in 200% hot water in which were introduced 300 g sulfuric acid 94% (representing 15% of the amount of sulfuric acid required for the reaction);

- glucose is dissolved in 200% water at 70-80°C.

Table 23: Reagent amounts used in the pilot scale synthesis of chromium-aluminum-iron heterocomplex solution.

Chromium-aluminum-iron heterocomplex $Cr_2O_3/Al_2O_3/Fe_2O_3$ (%)	Reagents used							
	Cr_2O_3 (g)	Sodium dichromate solution with concentration of 216g/l Cr_2O_3 (ml)	Al_2O_3 (%)	Aluminum sulfate solution with concentration of 53.4 g/l Al_2O_3 (%)	Fe_2O_3 (g)	Ferrous sulfate (g)	H_2SO_4 94% (g)	$C_6H_{12}O_6$ (g)
50/35/15	1100	5000	770	14400	330	1155	2000	770

Sodium dichromate solution is added to aluminum sulfate solution while continuously stirring. The reaction mass thus obtained is heated at 80°C, then the remaining sulfuric acid (1.7kg) and ferrous sulphate solution previously heated to about 80°C are added in a thin stream while continuously stirring.

After adding the ferrous sulphate solution, the reaction mass becomes brown. The glucose solution heated to 80°C is poured over it in a thin stream, continuously stirring. Synthesis reaction takes place slowly and is not violent due to dilution of solutions used. After the completion of the effervescent phase, produced by removing carbon dioxide, boiling continues for 30-60 minutes. The end of the reaction is checked by iodometric test.

The next day, the solution thus obtained is buffered with a solution of sodium bicarbonate of 5% concentration. The amount of baking soda used is about 14.7% of the metal oxide quantity. Analytical determinations for the solution prepared are given in Table **24**.

Table 24: Chemical characteristics for the chromium-aluminum-iron heterocomplex solution.

Chemical characteristics	MU	Values
Metal oxides	g/l	81
Cr_2O_3	g/l	43.07
Al_2O_3	g/l	26.18
Fe_2O_3	g/l	11.75
Basicity	%	18.48
pH as such		2.60
pH analytical solution		3.23

The stability of the solution obtained is expressed in precipitation points and pH values at these points, which are given in Table **25**.

Table 25: Precipitation points and pH values at precipitation points for the chromium-aluminum-iron solution.

Precipitation points with NaOH solution 0.5n initial final	0.8ml pH=3.41 8ml pH=3.98
Precipitation points with Na_2CO_3 solution 0.5n initial final	0.6ml pH=3.75 2.2ml pH=4.0
Precipitation points with $NaHCO_3$ solution 0.2n initial final	0.8ml pH=3.91 3.2ml pH=4.01

The chromium-aluminum-iron heterocomplex solution thus obtained is filtered and heated to 40-50°C, and then it is atomized.

Atomization process parameters are: inlet temperature 250-300°C, outlet temperature of 120-125°C and air pressure of 4 atmospheres.

Atomization leads to obtaining a fine, light green colored powder whose characteristics are presented in Table **26**.

Table 26: Chemical characteristics for chromium-aluminum-iron heterocomplexes in the form of powder.

Chemical characteristics	MU	Values
Moisture	%	4.22
Metal oxides	%	27.4*
Cr_2O_3	%	13.5*
Al_2O_3	%	9.78*
Fe_2O_3	%	4.1*
Basicity	%	20.8
pH solution 1:10		2.96
pH analytical solution		3.05

*recalculated for moisture-free

Table 27: Precipitation points and corresponding pH values for the chromium-aluminum-iron heterocomplex solution obtained from powder.

Precipitation points with NaOH 0.5n initial final	3.5ml pH=3.38 26.1 pH=3.89
Precipitation points with $NaHCO_3$ 0.5n initial final	11.5ml pH=4.06 11.5ml pH=4.06
Precipitation points with Na_2CO_3 0.2n initial final	11.5 pH=4.06 11.05 pH=4.06

For solutions obtained by dissolving powder chromium-aluminum-iron heterocomplexes, precipitation points at treatment with different alkalis and corresponding pH values are determined according to Table **27**.

In the case of chromium-aluminum-iron heterocomplexes, atomization does not lead to increased stability of the complex. We assume that the presence of aluminum is a factor of instability for the chromium-iron complex that does not allow penetration of sulfate groups in the complex, probably due to coordinative saturation or steric prevention.

Pilot Scale Synthesis of Chromium-Iron-Zirconium Tanning Metallic Heterocomplexes

To check pilot scale synthesis technology the product containing 70% Cr_2O_3, 10% Fe_2O_3 and 20% ZrO_2 was selected.

In order to obtain a solution containing about 2200 g metal oxides in the proportions shown above, the quantities of reagents shown in Table **28** were used.

Table 28: Reagents used in the pilot scale synthesis of chromium-iron-zirconium solution.

Cr/Fe/Zr hetero-complex solution (%)	Cr_2O_3 (g)	Sodium dichromate solution with 223g/l Cr_2O_3 (ml)	Fe_2O_3 (g)	Ferrous sulfate (g)	ZrO_2 (g)	Zirconium sulfate solution with 91.2g/l Zr (ml)	H_2SO_4 85% (g)	$C_6H_{12}O_6$ (g)
				Reagents used				
70/10/20	1540	6900	220	770	440	3570	1500	1078

The method used to obtain the heterocomplex solution based on chromium-iron-zirconium is described below:

- 85% of sulfuric acid (1.275 g sulfuric acid) is added to the sodium dichromate solution while continuously stirring, then the zirconium sulfate solution is added;

- ferrous sulfate is dissolved in 200% water acidified with 15% of the sulfuric acid amount required for the reaction;

- the acidified solution of ferrous sulphate is added to the system comprising sodium dichromate, sulfuric acid and zirconium sulfate while continuously stirring;

- the newly obtained system is heated to approximately 80°C. The heated glucose solution, obtained by dissolving sugar in 200% water, is added in thin stream, continuously stirring.

The reaction releases heat and abundant foam throughout the addition of glucose. The reduction is finalized after boiling the solution for about 60-80 minutes. The end of the reduction is checked by iodometric reaction.

The chemical characteristics of the chromium-iron-zirconium heterocomplex solution thus obtained are given in Table **29**.

Table 29: Chemical characteristics for the chromium-iron-zirconium heterocomplex solution.

Chemical characteristics	MU	Value
Metal oxides	g/l	168.12
Cr_2O_3	g/l	115.43
$Fe_2O_3 + ZrO_2$	g/l	52.69
Basicity	%	41.4
pH of solution as such		3.11
pH of analytical solution		3.48

Precipitation points and pH values determined at precipitation points for the chromium-iron-zirconium heterocomplex solution are given in Table **30**.

The solution is prepared for drying by filtering and preheating at 40-50°C. Atomization drying parameters are: inlet temperature of 250-300°C, outlet temperature of 100-120°C and air pressure of 4 atmospheres.

Chemical characteristics of the powder obtained from synthesized chromium-iron-zirconium heterocomplex solution are shown in Table **31**.

Table 30: Precipitation points and pH at precipitation points for the chromium-iron-zirconium heterocomplex solution.

Precipitation points with NaOH 0.5n: initial	8ml pH=5.09
final	8ml pH=5.09
Precipitation points with NaHCO$_3$ 0.5n initial	7ml pH=5.60
final	8ml pH=5.60
Precipitation points with Na$_2$CO$_3$ 0.2n initial	3.9ml pH=5.47
final	3.9ml pH=5.47

Table 31: Chemical characteristics of chromium-iron-zirconium heterocomplex powder.

Chemical characteristics	**MU**	**Value**
Moisture	%	2.65
Metal oxides	%	33.60*
Cr$_2$O$_3$	%	20.67*
Fe$_2$O$_3$+ ZrO$_2$	%	12.93*
Basicity	%	41
pH of solution 1:10		2.99
pH of analytical solution		3.23

*recalculated for moisture-free

Determinations for precipitation points and corresponding pH values, performed for solutions obtained by dissolving powder chromium-iron-zirconium heterocomplexes are presented in Table **32**.

By comparing the data in Table **30** with those in Table **32** it can be seen that atomization does not change the stability of complexes for the chromium-iron-zirconium heterocomplex. We assume that for the mixed complexes containing three types of metals, an additional anionization by atomization does not occur, probably because of the same steric prevention or coordinative saturation, as in the case of chromium-aluminum-iron complexes.

Table 32: Precipitation points and pH at precipitation points for the chromium-iron-zirconium heterocomplex solutions obtained from powder.

Precipitation points with NaOH 0.5n initial final	12.1ml pH=5.18 12.1ml pH=5.18
Precipitation points with Na_2CO_3 0.5n initial final	11.6ml pH=5.30 11.6ml pH=5.30
Precipitation points with $NaHCO_3$ 0.2n initial final	8.4ml pH=5.47 8.4ml pH=5.47

Pilot scale synthesis of tanning metallic heterocomplexes of chromium-iron, chromium-aluminum-iron and chromium-iron-zirconium aimed at verifying through practical experiments the synthesis technologies established under laboratory conditions and of the principles outlined in the chapter of theoretical study of synthesis mechanisms. Tanning metallic heterocomplexes can be synthesized according to theoretical principles established and work methods tested in the laboratory. The chemical properties of synthesized tanning metallic heterocomplex solutions are very similar to those obtained in the laboratory and indicate the possibility of their use in specific working conditions of leather tanning and retanning. Tanning metallic heterocomplex solutions can be processed and synthesized in the form of powders by atomization, a stable form that allows preservation of initial tanning properties and is suitable for addressing instrumental methods of structural analysis.

Stability of tanning chromium-iron heterocomplex powder to the action of alkali is improved compared with that of solutions, indicating masking with sulfate groups or organic residues by atomization.

Stability of tanning chromium-aluminum-iron and chromium-iron-zirconium heterocomplex powder to the action of alkali is unchanged compared with stability of heterocomplex solutions, which indicates a coordinative saturation in this case.

Send Orders for Reprints on reprints@benthamscience.net
Applicative Chemistry of Tanning Metallic Heterocomplexes, 2013, 95-122 **95**

CHAPTER 5

Structural Analysis of Tanning Metallic Heterocomplexes and Testing their Tanning Properties

Abstract: Knowing the structure of tanning metallic heterocomplexes is closely linked to their chemical properties in relation to stability in various environments specific for natural leather tanning and to their ionic nature which influences their penetration in the dermis of natural leathers. This chapter deals with specific analyses for coordination complexes, for the purpose of anticipating the behaviour of tanning metallic heterocomplexes upon the interaction with natural leather. IR and electron spectroscopy have enabled to identify organic ligands, the polynuclearity of complexes by means of sulphate or hydroxyl anions, the metal-oxygen-metal interactions and have eliminated the hypothesis of direct metal-metal interactions. Electron spin resonance spectroscopy allowed identification of structures in predominantly octahedral environments with rare insertions of tetrahedral environments, which provides steric stability to heterocomplex structures. SEM-EDAX analysis determines the existence of tanning metals used and morphological particularities at microscopic level. Analysis of ionic components explains the more pregnant anionic nature compared to that of the best known tanning agent, basic chromium salt, and therefore, the improved efficiency of using tanning metallic heterocomplexes in natural leather tanning.

Keywords: IR spectroscopy, Electron spectroscopy, Electron spin resonance spectroscopy, SEM-EDAX, Ion exchange chromatography, Tanning properties.

STRUCTURAL CHARACTERISTICS OF TANNING METALLIC HETEROCOMPLEXES

In order to understand important structural elements of synthesized tanning metallic heterocomplexes, such as: the type of ligands, the type of chemical bonds, stereochemistry, oxidation state of chromium and iron, powder composition and morphology, a number of methods of investigation specific to coordination chemistry are used, such as IR vibrational spectroscopy, reflection electron spectroscopy (RES), scanning electron microscopy and energy-dispersive X-ray analysis. These analyses were performed on tanning metallic heterocomplexes brought to solid form by atomization, the same technique for basic chromium salt preparation.

Also, in order to better explain the interaction of tanning metallic heterocomplexes with collagen, chromatographic separation of polydisperse

Carmen Gaidau

components in solutions was performed and the crosslinking ability of each component was determined.

Infrared spectroscopy is a method with high selectivity and quick response, which provides valuable information on how various ligands are coordinated, on the stereostructure of coordination compounds and sometimes on the strength of the covalent bonds. In the vibrational spectra of coordination compounds several types of vibrations can be identified due to: the ligand, which recall those of the free ligand; the lattice that characterizes the entire molecule; the vibration coupling of two ligands or of the lattice with one of the ligands (coupled vibrations).

Given the literature reports [116-124] and the spectra obtained for the studied substances (Figs. **18, 19** and **20**), wave numbers of absorption peaks in IR spectra, qualitative characterization of the shape and intensity of bands extracted with assignments extracted from the listed sources were tabulated (Table **33**).

The study of polynuclear complex combinations by vibrational spectroscopy aimed at analyzing modes of vibration of ligands acting as bridges. Types of bridges identified in medium and strong bands were: OH, corresponding to strong broad bands, with peaks ranging between 3500 cm^{-1} and 3200 cm^{-1}; SO_4, for which the split of bands in the 1000-1200 cm^{-1} region, specific to vibrational mode υ_3, indicates the presence of a deformed symmetry c_{2v}, typical for a bridge group; Me-O, SO_4, in the case of heteronuclear species, the interaction between different metals and intermediate groups in the coordination sphere results in a broad band between 650-400 cm^{-1}, which is complex and in which components mainly due to Me-O bond and υ_2 frequencies of SO_4 ion can sometimes be distinguished, and sometimes not and aldehydic CH, CO, CH_2, highlighted by the occurrence of υ_{CH} vibrations in CH_2 groups, of υ_{CO} valence vibrations and which lead to the conclusion of the existence of compounds specific to glucose decomposition. Of course, the presence of broad absorption band at 500 cm^{-1} frequency is highlighted, specific to chromium.

Table 33: Characteristic IR absorption frequencies of chromium-iron, chromium-aluminum-iron and chromium-iron-zirconium heterocomplexes.

Cr-Fe υ (cm^{-1}) (Intensity*)	Cr-Al-Fe υ (cm^{-1}) (Intensity*)	Cr-Fe-Zr υ (cm^{-1}) (Intensity*)	Assignment
3340 (m)	3280 (s)	3355 (w) 3275 (w)	υ_{OH} by hydrogen bridges
2905 (m) 2855 (m)	2945 (s) 2880 (s)	2905 (m) 2850 (m)	υ_{CH} in aldehydes
1740 (w) 1680 (m) 1625 (m)	2500 (w) 2350 (w) 1660 (m)	1740 (w) 1700 (w) 1620 (w)	υ_{CO} asymmetrical υ_{CO} symmetrical δ_{HOH}
1600 (m) 1535 (w)	1500 (w)	1590 (w) 1500 (w)	
1450 (w)	1420 (w)	1440 (w)	δ_{CH} in CH$_2$
1420 (w)		1400 (w)	
1180 (s) 1080 (s) 1045 (s)	1180 (s) 1080 (s) 1050 (s)	1180 (s) 1080 (s) 1040 (s)	υ_3 for SO$_4$ in symmetry c$_{2v}$ deformation and δ_{CO}
860 (w) 640 (m)		840 (w) 640 (m)	
	860 (m) 600 (m)		δ_{OSO} deformation
520 (w)		500 (m)	υ_{Cr-O}=457 cm^{-1}; $\upsilon_{2\,SO4}$=460 cm^{-1} υ_{Fe-O}=434 cm^{-1}
	520 (m) 480 (m)		Overlapping of: υ_{Al-O}=490 cm^{-1}; υ_{Cr-O}=459 cm^{-1};υ_{Fe-O}=434 cm^{-1}; $\upsilon_{2\,SO4}$=460 cm^{-1}

*Intensity: s-strong, m-medium, w-weak.

In Table **33**, all frequencies which may overlap in this domain are mentioned in the assignments column according to [116]. The common feature for all trinuclear combinations (chromium-aluminum-iron and chromium-iron-zirconium) is the complex appearance of the band located at the lower end of the work domain due to overlapping of a large number of possible frequencies.

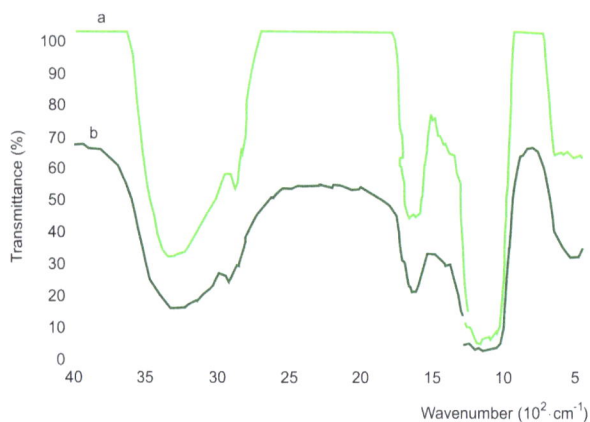

Figure 18: IR spectra for Cr-Fe heterocomplex (without-a and with-b attenuation).

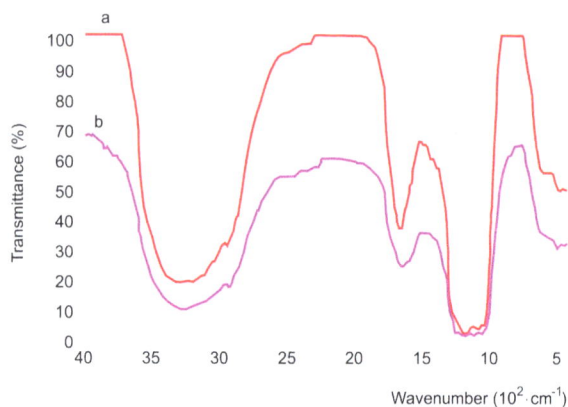

Figure 19: IR spectra for Cr-Al-Fe heterocomplex (without-a and with-b attenuation).

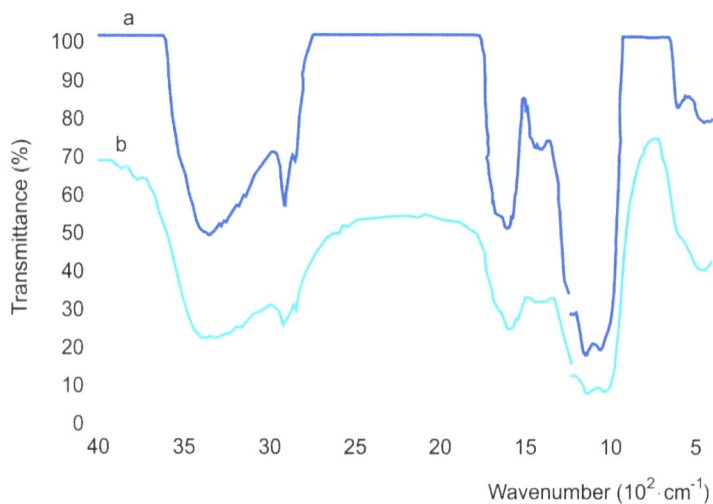

Figure 20: IR spectra for Cr-Fe-Zr heterocomplex (without-a and with-b attenuation).

Near Infrared Spectroscopy (200-400 cm^{-1})

This type of analysis was performed in order to elucidate the assumption that the relatively high stability of tanning metallic heterocomplexes can be explained by the existence of special metal-metal bonds.

A large number of Cotton's [124] research works in the field of coordination chemistry are devoted to multiple metal-metal bonds. The existence of such bonds has been proven; it is interesting that the first compound discovered containing a quadruple metal-metal bond was chromium dimer with the formula $[Cr_2(OOCCH_3)_4(H_2O)_2]$ [125,126].

As it is known, the factors on which metal-metal bond formation depends are: the number of valence electrons, the type of radial expansion of orbitals used for metal-metal bond formation, the oxidation state of the metal ion and, the nature of ligands.

The nature of ligands is a very important factor that influences the formation of metal-metal bonds because it determines the formal oxidation state of the metal ion and the radial extension of their orbitals. Given the complexity of the reaction systems forming during synthesis, these assumptions were testes by near IR spectra measurements (200-400 cm^{-1}) - specific domain for metal-metal bonds of chromium-iron heterocomplex (Fig. **21**).

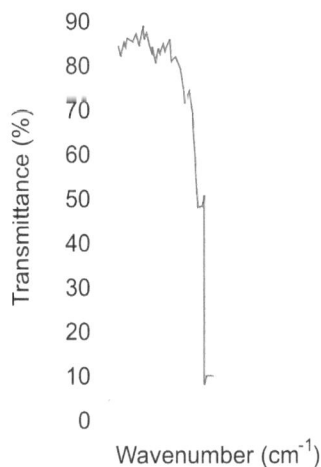

Figure 21: IR spectrum in the near domain for chromium-iron heterocomplex.

Whether certain atoms or groups of atoms act as a bridge between metal ions is well established by IR analysis.

The number of IR active vibrational modes is dependent on the symmetry of that group as well as on the type of bridge it makes. Thus, the M-O-M system can be linear or angular.

Frequencies of vibration of this M-O-M lattice do not depend on the atomic masses of metal ions, which means that among the metal-oxygen bonds, the elongation force constants are not affected by the atomic mass of M, much higher than that of the oxygen atom in the bridge.

In the case of angular M-O-M systems it can be foreseen that υ_{sim} frequencies (found between 200 cm^{-1} and 300 cm^{-1}) and υ_{asim} frequencies (found between 770 cm^{-1} and 880 cm^{-1}) will have higher and lower values, respectively, than in the case of linear M-O-M lattice.

The explanation is that, for angular systems, the π nature of the M-O bond is lower than that of linear systems, in which, for geometric reasons, d_{π}-p_{π} coverage is greater.

According to Table **34**, the assigned frequencies of vibration do not correspond to metal-metal bonds. Identified frequencies are higher than the frequencies corresponding to metal-metal bonds, which are: $\upsilon_{Cr-Cr} = 220$ cm^{-1} and $\upsilon_{Cr-Fe} = 232$ cm^{-1}.

Table 34: IR vibration frequencies (200-400cm^{-1}) from spectra of chromium-iron heterocomplex and probable assignments.

Sample	Vibration frequencies (cm^{-1})	Assignments
Cr-Fe	270	υsim Cr-O-Fe
	335	υsim Fe-OH-Fe

This observation leads to the conclusion that the polynuclearity of these complexes is not achieved by metal-metal bonds, but by ligand bridges (SO_4^{2-}, OH$^-$). Montgomery [127] reached the same conclusion, formulating the heteronuclear chromium-zirconium complex without metal-metal bonds and only with OH^- and SO_4^{2-} bridges.

Reflection Spectroscopy

UV-VIS spectroscopy in solvent has some serious limitations due to the influence of the solvent, so in the chemistry of complex combinations, more accurate information about the stereochemistry of a complex combination is provided by spectral methods recorded on crystalline powders (solid samples diluted with MgO), the so-called reflection electronic spectra.

In the visible domain, complex combinations containing ions with d^n configurations provide important data on the central metal coordination.

In the visible and UV domain a complex combination presents several absorption bands very different in intensity and form, having different origins:

- bands assigned to the ligand field (in the visible and ultraviolet domain);

- bands assigned to oxidation-reduction process (bands with charge transfer in the ultraviolet domain);

- bands of the ligand (in the ultraviolet domain).

The characteristic structure for complex combinations of Cr (III) is generally the octahedral hexacoordinated one. The number of complex combinations where this ion is found in a tetrahedral environment is extremely low (only two such combinations are reported, $(PtCl_4)$ $(CrCl_4)$ and chromium-wolframic acid).

The Tanabe-Sugano diagram of metal ions with d^3 configuration in an octahedral field (such as Cr III), provides three spin-allowed transitions:

$$^4T_{2g}(P) \longleftarrow ^4A_{2g}$$

$$^4T_{1g}(F) \longleftarrow ^4A_{2g}$$

$$^4T_{1g}(P) \longleftarrow ^4A_{2g}$$

According to these transitions, three absorption bands in the high wavelength range are expected.

In the electron spectra of complex combinations for which ligands do not have absorption in the UV domain, all three bands assigned to above-mentioned transitions can be seen.

In general, however, for most complex combinations, the high energy band (which occurs at about 30 000 cm^{-1}) and the low intensity band assigned to the $^4T_{1g}(P) < ---$ --- A_{2g} transition is covered by charge transfer bands or by the ligand bands.

Under certain conditions bands assigned to spin-forbidden transitions were observed.

In the spectra of hexacoordinate complex combinations of Cr (III) of reduced symmetry, the band assigned to spin-allowed transition $^4T_{2g}(P) <----- \ ^4A_{2g}$, corresponding to the lowest energy, often has a prominent split, which was attributed to the field decrease or to σ and π contributions of the ligands.

The recording of electronic spectra of chromium-iron, chromium-aluminum-iron and chromium-iron-zirconium heterocomplexes compared to the chromium complex is shown in Fig. **22**. Assignments of wavelength variation are given in Table **35**.

The appearance of electron spectra of tanning metallic heterocomplexes is slightly changed compared to the classic chromium complex spectrum. It can be assumed that this change is due to the presence of Fe(III)-d^5, which presents a distorted environment. Bands in the visible (d-d) domain of the spectrum of iron complexes are generally not very strong. According to the Tanabe-Sugano diagram, for a d^5 ion, spectral lines of iron come from spin-forbidden transitions which become spin-allowed as a result of mixing sextet states to quartet states by spin-orbit coupling.

The d-d spectra of high-spin complexes of Fe (III) are generally more difficult to assign. The presence of wave number in the 750-800 nm domain in spectra of samples containing Fe (III) in addition to Cr (III) can be considered as an indication of the existence of octahedral Fe (III) in these circumstances.

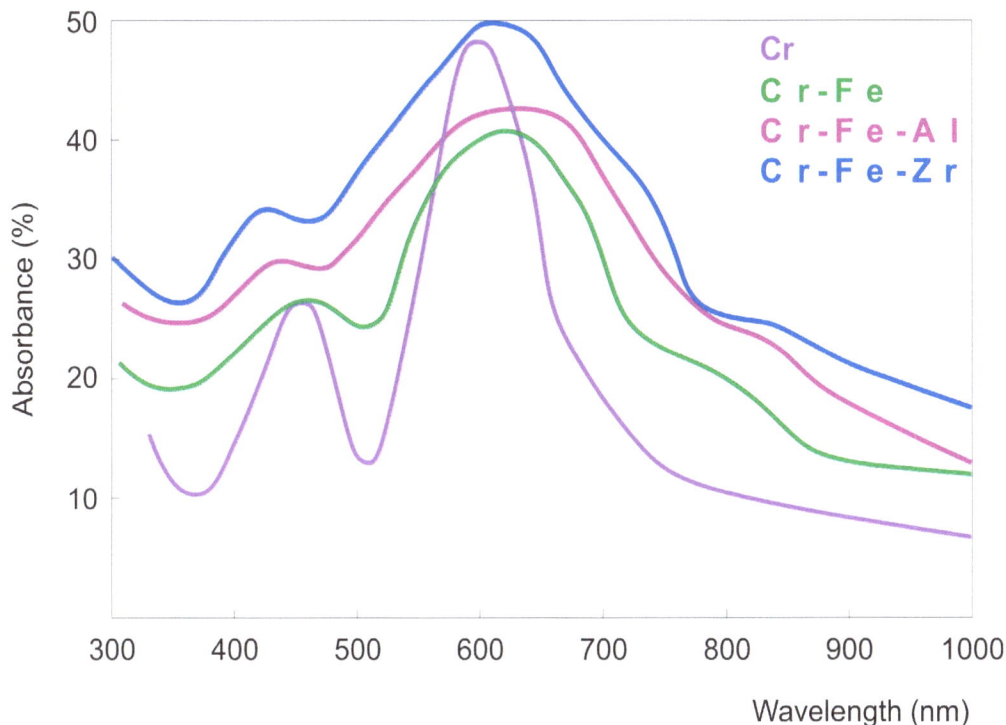

Figure 22: Reflection electron spectra for tanning metallic heterocomplexes of chromium-iron, chromium-aluminum-iron and chromium-iron-zirconium.

Chromium-iron complex spectrum is similar to that of chromium-aluminum-iron and chromium-iron-zirconium complexes.

Based on reflection electron spectroscopy data, it can be concluded that heterocomplexes containing $CrIII(d^3)$ and $FeIII(d^5)$ are in a weakly distorted octahedral environments, very likely due to the non-equivalence of ligands (OH⁻, SO_4^{2-} groups, water molecules *etc.*).

The presence of metals which do not absorb in the visible domain (aluminum and zirconium) does not generally affect the reflection spectra shape.

The spectroscopy analyses performed demonstrated that the polynuclearity of the three tanning metallic heterocomplexes studied is achieved by: SO^{2-}, OH⁻ bridges;

Cr-O, Fe-O, Al-O bonds and aldehyde-type ligands derived from decomposition of organic reductant.

Table 35: Reflection electron spectra (300-1100 nm) of chromium-iron, chromium-aluminum-iron and chromium-iron-zirconium heterocomplex samples.

Sample	Bands (nm)	Assignment	Stereochemistry [116-118]
Cr-Fe	330 (CT) 420 565 (asym) 750 (shoulder) (spin-forbidden)	$^4T_{2g}(P) \longleftarrow ^4A_{2g}$ $^4T_{1g}(F) \longleftarrow ^4A_{2g}$ $^4T_{2g}(F) \longleftarrow ^4A_{2g}$ $^4T_{1g} \longleftarrow ^6A_{1g}$	Cr^{3+} in tetragonal distorted Oh environment Fe^{3+} in Oh environment
Cr-Al-Fe	330 (CT) 425 570 (asym) 770 (shoulder) (spin-forbidden)	$^4T_{2g}(P) \longleftarrow ^4A_{2g}$ $^4T_{1g}(F) \longleftarrow ^4A_{2g}$ $^4T_{2g}(F) \longleftarrow ^4A_{2g}$ $^4T_{1g} \longleftarrow ^6A_{1g}$	Cr^{3+} in tetragonal distorted Oh environment Fe^{3+} in Oh environment
Cr-Fe-Zr	330 (CT) 420 570 (asym) 760 (shoulder) (spin-forbidden)	$^4T_{2g}(P) \longleftarrow ^4A_{2g}$ $^4T_{1g}(F) \longleftarrow ^4A_{2g}$ $^4T_{2g}(F) \longleftarrow ^4A_{2g}$ $^4T_{1g} \longleftarrow ^6A_{1g}$	Cr^{3+} in tetragonal distorted Oh environment Fe^{3+} in Oh environment

Electron Paramagnetic Resonance (EPR)

EPR is a spectroscopic study method that uses electromagnetic radiation in the microwave range to simulate electronic transitions occurring between various energy levels defined by the orientation of electronic magnetic moments in an external magnetic field. These levels are partially occupied at room temperature and represent the group of lowest energy levels (basic) of the paramagnetic center located in a magnetic field.

Information that can be obtained from EPR spectroscopy applied to the study of materials are related to the three essential characteristics of a paramagnetic system:

a) effective magnetic moment of the unpaired electron (electronic g factors);

b) interaction between the electronic magnetic moment and its vicinity, which allows the electronic spin system to exchange energy with its vicinity and thus maintain thermal equilibrium between electronic spin energy levels (spin-lattice relaxation) and

c) interactions between the electronic magnetic moment and the different nuclear magnetic moments in the system (hyperfine interactions) that provide the so-called A factors.

The g and A factors are the most useful features that can be obtained from an EPR spectrum. Hyperfine interactions are related to unpaired electron load distribution in the basic state, while electronic g factors are related to the nature of excited states, which can be coupled to the basic state by electronic angular moment components along the different molecular axes.

The essential characteristics of a resonance signal are:

-   resonance field, BO, allowing calculation of g factor, knowing the frequency of the microwave field (g=hυ/BBO).

-   resonance line width, ΔB, measured between the points of maximum inflection of the resonance curve (the derived curve). Line broadening mechanism, as a function of a number of physical factors, may be a feature of particular importance;

-   resonance line profile can be simple (such as the Lorenzian profile of homogeneously broadened curve) or Gaussian (inhomogeneous broadening due to static variations in the local magnetic field). In many cases the profile is not simple, and the changing line shape can provide information on the interactions of paramagnetic centers;

-   the area under the absorption curve, which is proportional to the number of paramagnetic centers in the sample and can be calculated by the graphic integration of the curve area and compared with a standard sample showing a similar profile line.

The usual procedure of a resonance study consists in analyzing the resonance line spectrum and determining the values of the main constants of various interactions, and then linking the observed values of these constants (molecular) to the system structure.

Spin resonance measurements indicate the presence of significant signals in all samples. These signals were analyzed and the main parameters of resonance lines were extracted. Recorded spin resonance parameters are presented in Table **36** and Fig. **23**.

Table 36: Recorded spin resonance parameters for chromium-iron, chromium-aluminum-iron and chromium-iron-zirconium heterocomplexes compared to those of chromium complex.

Sample	g	B (mT)	I Δ B# 2 (arbitrary unit)	Observation
Cr	1.980	55	32.6	
Cr-Fe	1.980	75	36	Changed shape
Cr-Al-Fe	1.981	44	33.6	Changed shape
Cr-Fe-Zr	1.980	82	36.3	Changed shape

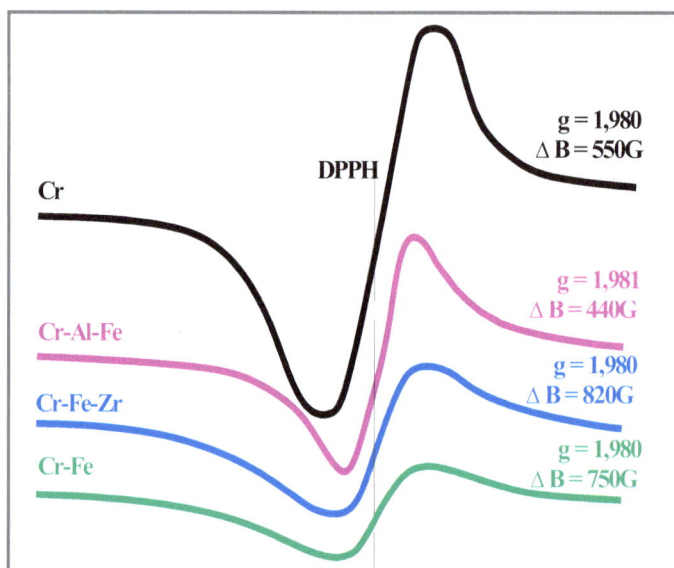

Figure 23: Electron spin resonance spectrum for the chromium complex and chromium-iron, chromium-aluminum-iron and chromium-iron-zirconium heterocomplexes.

The shape of resonance signal observed for the chromium-iron heterocomplex is significantly different from the signal emitted by the reference sample - chromium complex, which is attributed to the presence of iron in the vicinity of chromium.

The composite shape of the resonance signal indicates that we are dealing with two paramagnetic species. It is possible that in this case the isolated vicinity of iron could cause the emergence of a new paramagnetic species.

EPR analysis results indicate that in the chromium-iron heterocomplex samples, iron octahedra bind preferentially through peaks and we are dealing with CrO_6 and FeO_6 configurations with weaker interactions than for material compaction with octahedra ordering by the edges.

The chromium-aluminum-iron heterocomplex sample is very similar in terms of the EPR signal to that of chromium-iron.

For this sample also it is confirmed that iron is the element that influences the Cr^{3+} signal and in the case of this heterocomplex, a double signal is recorded, which indicates the presence of two paramagnetic species based on chromium and iron.

The presence of a very weak signal in the resonance spectrum, located at a field of B=1500 GS is also noted, which can be attributed to Fe^{3+} in tetrahedral environment.

In the case of chromium-iron-zirconium heterocomplex the presence of a strong resonance signal can be noticed, which is very close to the chromium-iron heterocomplex sample signal. At the same time there is a change of the resonance line shape.

By comparing the three EPS curves, it can be stated that aluminum and zirconium affect less the structure of heterocomplexes and it seems that the role of iron has a greater importance; it governs interaction with Cr^{3+} ions.

Structural determinations confirm the presence of iron ions, Fe^{3+} with octahedral environment, that lend themselves to mutual ordering with Cr^{3+} ions, to which they are related.

Scanning Electron Microscopy (SEM) and Energy-Dispersive X-Ray Analysis (EDAX)

This type of analysis provides data on morphology and aggregation state of the particles using the modern technology of scanning electron microscopy. Since all samples are of dielectric nature, they can be covered with a metal layer of copper. In addition, we monitored distribution of relevant elements in samples (heavy elements) using EDAX technique (energy-dispersive X-ray analysis). Images shown in Figs. **24, 25** and **26** show that chromium-iron heterocomplex powders have coarse grains of irregular morphology, with rounded edges. Few small particles keep their quasi-spherical morphology; they have dimensions of 10-15 μm.

Chromium distribution was relatively homogeneous and iron is mainly concentrated in non-spherical particles. Chromium-aluminum-iron heterocomplex powder appears as a compact, glassy, typical amorphous mass crossed by a network of cracks and chromium distribution is relatively homogeneous.

Chromium-iron-zirconium heterocomplex powder has a mainly spherical morphology, with sphere diameter of 5-25 μm, with fragments of broken spheres and a certain concentration of components with vitreous aspect, with sizes between 20-50 μm. Chromium is uniformly distributed in particles, in high concentration. Iron distribution follows that of chromium, and zirconium is slightly above the detection limit.

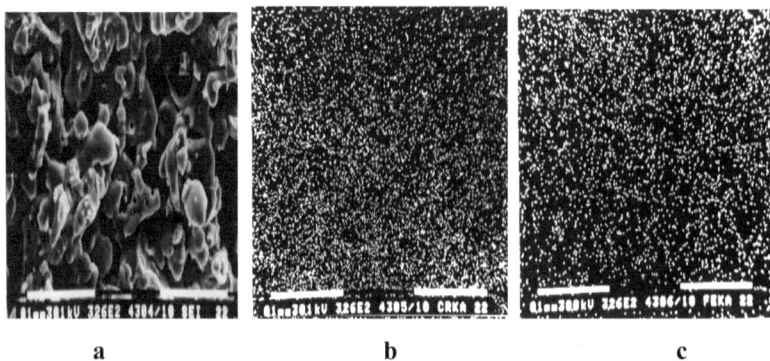

a   b   c

Figure 24: a.Morphology of chromium-iron heterocomplex particles; b.Distribution of chromium in sample; c. Distribution of iron in sample.

a

b

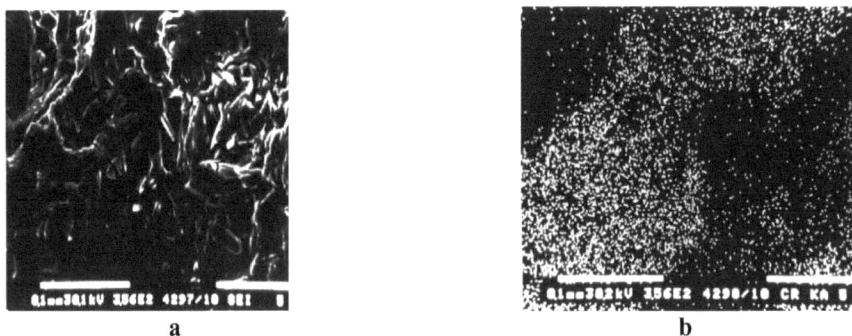

Figure 25:-a.Morphology of chromium-aluminum-iron heterocomplex particles; b.Distribution of chromium in sample.

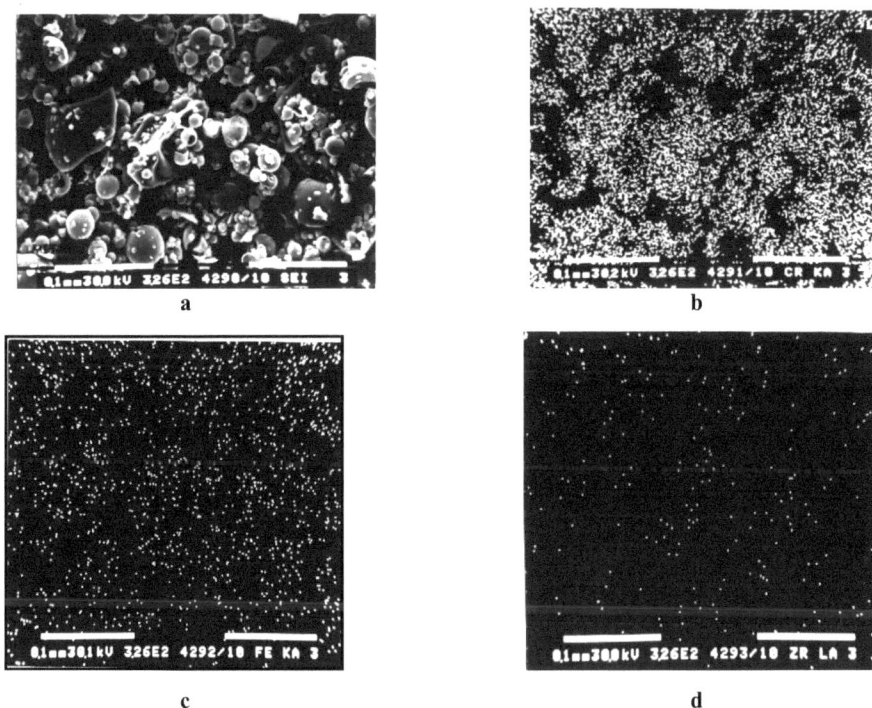

a

b

c

d

Figure 26: a. Morphology of chromium-iron-zirconium heterocomplex particles; b. Distribution of chromium in sample; c. Distribution of iron in sample; d. Distribution of zirconium in sample.

Analyses on Metal Heterocomplex Solutions

As it is known, tanning solutions of metal complexes are polydisperse solutions whose tanning ability lies in the many possibilities to form cross-links in the collagen molecule structure.

This polydisperse composition was researched with greater accuracy [128] by using methods of Sephadex gel separation and IR analysis.

Identifying the types of complexes, their electrical loads, determining their ability to attach to the hide powder have led to the conclusion that they have different penetration abilities and the hypothesis of a selective combination with collagen was issued. Their mixture and polydisperse character ultimately provides the maximum tanning effect.

The study of synthesis mechanisms and the study of basicity evolution in tanning metallic heterocomplex solutions show their different behavior in comparison with the basic chromium solution.

The higher acidity of tanning metallic heterocomplexes leads to the hypothesis of a higher anionization in comparison with the basic chromium complexes with important influences in terms of penetration power, exhaustion of chromium from baths *etc.*

Performing a separation of tanning metallic heterocomplex solutions using the ion exchange separation technique and analysis of fractions obtained represents an original approach.

Determining the Electric Charge of Tanning Solution Components

To establish the electrical charge of the complexes ion exchange [129] is used: Dowex 1 is used for anions, whose active group is $-N^+(CH_3)_3$, and Dowex 50 for cations, whose active group is $-SO_3^{2-}$. The results obtained for the three types of tanning metallic heterocomplexes compared to basic chromium salt are presented in Table **37**.

Table 37: Electric charge of tanning metallic heterocomplexes compared to basic chromium salts.

Electric charge	% Cr of total Cr			
	Cr	Cr-Fe	Cr-Al-Fe	Cr-Fe-Zr
Cationic	73.4	53.6	46.2	61.7
Anionic	-	29.3	34.8	27.1
Neutral	26.6	17.1	19.0	11.2

The presence of a stable anionic fraction in tanning metallic heterocomplexes can be explained by the synthesis reaction mechanism, which requires large quantities of sulfuric acid and implies the existence of a larger quantity of SO_4^{2-} groups which can enter the coordination sphere and anionize complexes. Also, using a larger proportion of glucose may cause increased anionization with organic ligands.

The existence of an anionic fraction with higher amount of tanning metallic heterocomplexes compared to basic chromium complexes has important implications on the tanning properties of these new tanning materials. Thus, they must have a higher rate of hide penetration and if they also have good ability of locking to hide, it can be assumed that chromium exhaustion from the tanning bath is better.

In order to emphasize the polydisperse character of the tanning metallic heterocomplex solutions and tanning ability of components, another type of separation on Sephadex C-25 gel was used [130, 131].

Separating Components from the Chromium-Iron Tanning Solution. Chemical and Spectral Analysis of Components

In order to separate components of the chromium-iron heterocomplex solution, a solution with a Cr concentration of 0.1M and 0.05M, respectively, is used. Separation using Sephadex C-25 gel is achieved by gradient elution with NaCl solution in concentration of 3M and 1.5M respectively. Molecular extinction coefficients, ε, were recorded at $\lambda=420$nm (c=0.1M) and $\lambda=420$nm (c=0.05m). Chromatograms on Sephadex gel of the chromium and iron based tanning solution are shown in Fig. **27**.

Chromium-iron tanning solution of 0.1M concentration can be separated into four distinct components of different colors by elution with 3M NaCl solution; in this case, on the Sephadex column a reddish-brown amorphous component remains (amorphous precipitate), which cannot be eluted with 3M NaCl solution; it is assumed that such a component, probably a Fe (III) salt is formed in the concentrated solution of 3M NaCl under the influence of Sephadex hydroxyl groups.

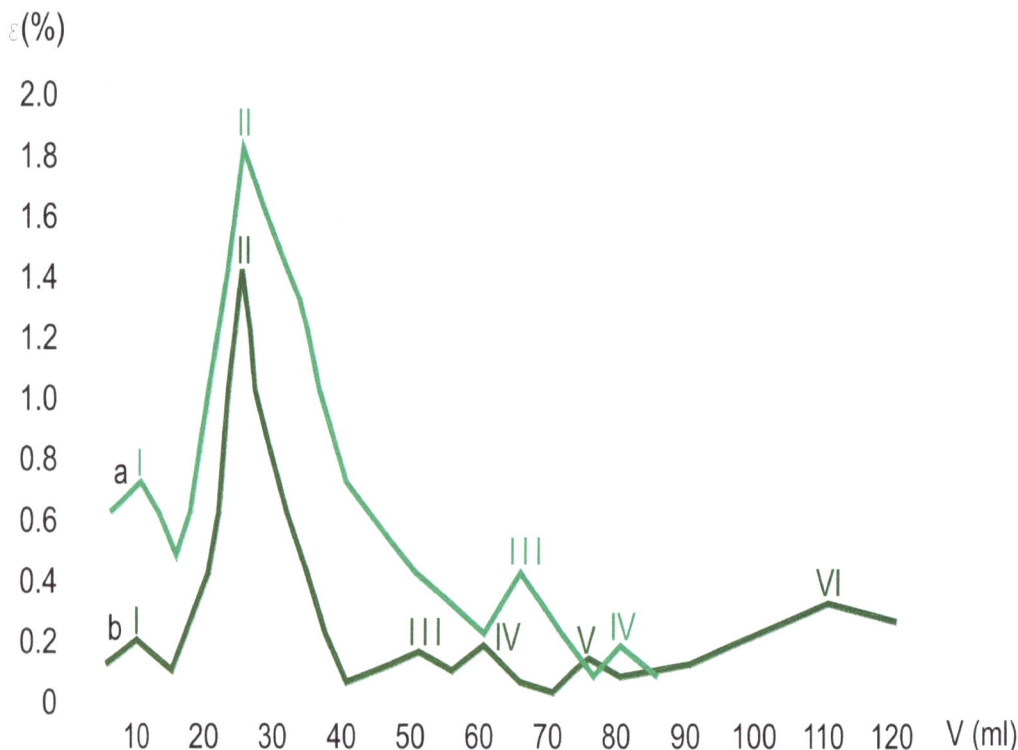

Figure 27: Chromatogram of the chromium-iron solution, c= 0.1M (a) and c= 0.05M (b) eluted on Sephadex C 25 gel.

The tanning solution in 0.05m concentration can be separated into six distinct components using 3M NaCl for elution; the last reddish-brown amorphous component, is fixed to the column and in this case as well, the component cannot be eluted with 3M NaCl solution, but was eluted with 0.1M HCI (40ml).

To avoid introducing 2 elution reagents (of which 0.1M HCl, which changes the composition of the eluted component) elution with 1.5M NaCl solution was attempted.The chromatogram of chromium-iron tanning solution in 0.1M concentration separated using Sephadex C 25 gel by elution with 1.5M NaCl is shown in Fig. **28**.

Five distinct components colored blue-green, blue, yellow-green, magenta and red were separated. These components were characterized by chemical analysis.

Chemical analysis data are presented in Tables **38** and **39**.

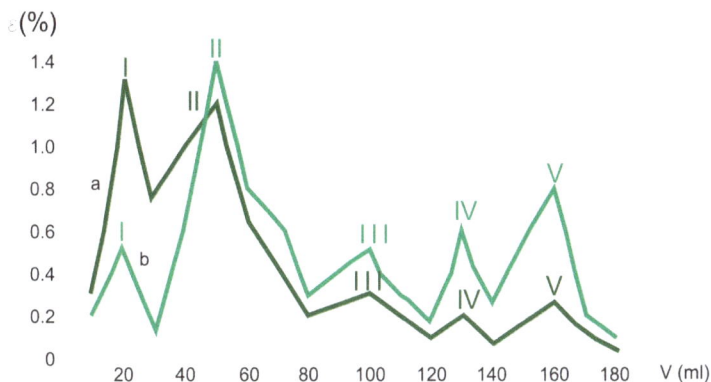

Figure 28: Chromatograms of the chromium-iron solution obtained by elution with a NaCl 1.5M solution, λ=420nm (a) and λ=580nm (b).

Table 38: Chemical analysis of components separated from the chromium-iron tanning solution using Sephadex C 25 (c=0.1M Cr).

Separated components	Cr g/l(moles)	Fe g/l(moles)	SO₄ g/l ionic coordinated		HCOO⁻ CH₃COO⁻ g/l(moles)	C₂O₄²⁻ g/l(moles)
I	1.03 (0.019)	-	3.53 (0.036)	-	-	-
II	0.97 (0.018)	-		0.69 (0.0072)	0.26 (0.006)	0.62 (0.0076)
III	1.53 (0.029)	-	1.41 (0.015)	-	-	1.30 (0.014)
IV	1.97 (0.037)	0.35 (0.006)	1.50 (0.015)	-	-	2.11 (0.024)
V	-	1.75 (0.031)	1.34	1.34 (0.014)	-	2.46 (0.028)

Table 39: Molar ratios of acid groups/M and M/M ratio in the components separated from the chromium-iron heterocomplex solution.

Separated component	SO₄²⁻/Cr	HCOO⁻/Cr	C₂O₄²⁻/M (M=Cr,Fe)	Cr/Fe
I	1.89	-	-	-
II	0.40	0.34	0.38	-
III	0.52	-	0.48	-
IV	0.41	-	0.65	0.16
V	-	-	0.93 (C₂O₄²⁻/Fe)	-

Next, a number of separate components were analyzed by UV-VIS and IR spectroscopy. For IR analysis, each component was converted to solid state by freezing and then KBr pelleting. UV-VIS spectra for components I, IV and V compared to the initial solution of chromium-iron heterocomplexes are shown in Fig. **29**.

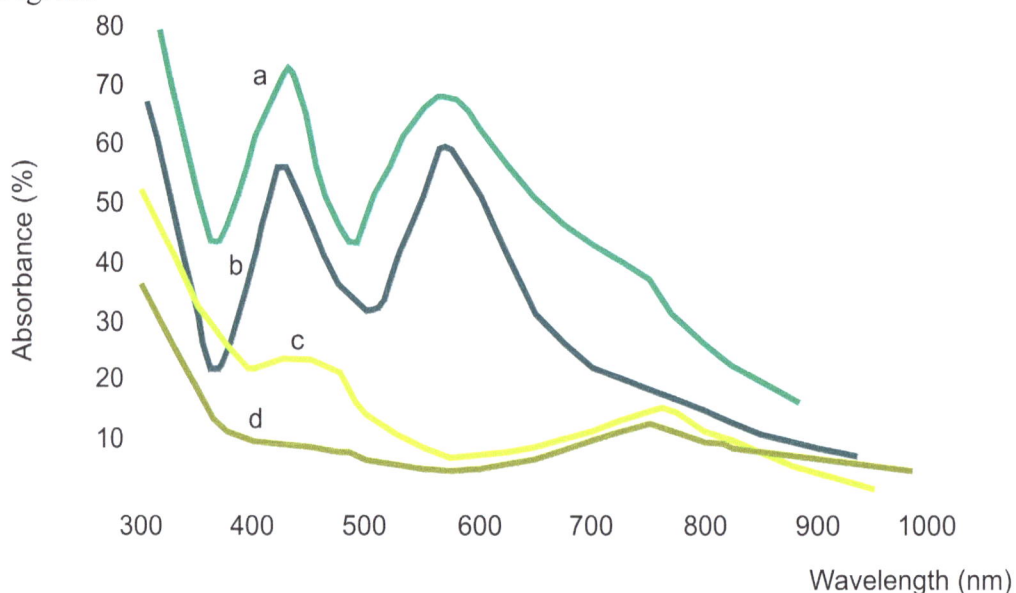

Figure 29: Electron spectra of components I (b), IV (c) and V (d) separated from the chromium-iron heterocomplex solution compared to the initial solution (a).

Electron spectra in the visible domain of components I, II and III are similar and reveal the presence of Cr (III) in a weakly distorted octahedral environment (peaks appear around 420 nm and 580 nm).

The electron spectrum of component IV is different from those of components I, II and III and of the initial solution; it presents a rudimentary band at 450 nm and a weak band at λ=750 nm, specific to Fe (III) in an octahedral environment.

Component V has a single characteristic band at about 750 cm^{-1} specific to Fe (III).

The main bands of IR spectra with probable assignments for the 5 components separated from the chromium-iron heterocomplex solution are presented in Table **40**. Fig. **30** presents IR spectra for components I, II and V.

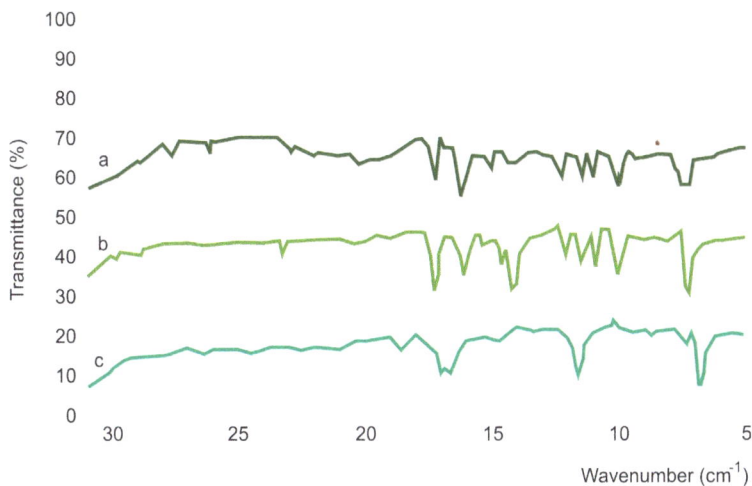

Figure 30: IR spectra for components I (a), II (b) and V (c) separated from the chromium-iron heterocomplex solution.

Table 40: The main bands of IR spectra and probable assignments for the chromium-iron solution components.

Component	SO_4^{2-}	$HCOO^-$	$C_2O_4^{2-}$	OH^-	H_2O
I	620(s) 1100(s)	-	-	-	1600(m) 3600(m)
II	640(m) 970(m) 1070(s) 1150(s)	1370(s) 1590(m)	1400(m) 1680(m)	950(m) 1100(w)	1620(m) 3500(m)
III	620 1050 1150	-	1410 1680	960 1100	1630 3500
IV	630 950 1160	-	1410 1660	960 1120	1630 3500
V	620 1110	-	1390 1670	930 1150	1610 3500

s-strong, m-medium, w-weak.

The spectrum of component I includes bands characteristic of ionic or coordinated SO_4^{2-} and frequencies characteristic of H_2O respectively.

The spectrum of component II includes bands characteristic of SO_4^{2-}, carboxylated anions ($HCOO^-$, $C_2O_4^{2-}$), and OH^- and H_2O coordination groups. The spectra of components III-V include bands characteristic to SO_4^{2-}, $C_2O_4^{2-}$, OH^- and H_2O anions.

Three components containing only chromium compounds, one component containing chromium-iron complexes and one which is probably a basic complex with organic iron ligands were separated using the chromatographic separation technique. The technique provides an imperfect picture of a probable composition because the basicity of ion exchanging gel, the dilution of the analyzed solution, and the interaction with the eluent are all elements disturbing the original structure of the initial solution. Separation studies are original for such mixed complexes; so far such analyses have only been done for homeopolynuclear complexes. It is very likely that the structure of tanning metallic heterocomplex solutions consists of a part of complexes specific to basic chromium solutions and a strongly anionized mixed complexed part.

We think that component V does not actually exist in this form in the chromium-iron heterocomplex (the initial solution does not have any reddish tint, not even in advanced dilution); it is a result of complex decomposition during operations specific to chromatographic technique.

However, we believe that during the leather processing, the processes that occur as a result of basification, neutralization *etc.* can be likened to the effect that the chromatographic column has on the chromium-iron complexes and a series of splits of the mixed compounds may occur in leather.

The tanning ability of the five components separated by chromatography was tested on pickled cattle hides calculating for each of the components an offer of 2% metal oxides in a classical tanning process. The results obtained are summarized in Table **41**.

It can be seen that none of the fractions tested has as good a tanning capacity as the initial solution. The results show that each component acts in a specific way, perhaps the final effect is synergistic.

Table 41: Shrinkage temperatures for cattle hides tanned with components separated from the chromium-iron heterocomplex solution, compared to the initial solution.

Sample	Tanning solution concentration (g/l)		Metal oxide offer (%)	Shrinkage temperature ($^{\circ}$C)
	Cr_2O_3	Fe_2O_3		
Fraction I	8.75	-	2	85
Fraction II	6.95	-	2	80
Fraction IV	6.1	1.5	2	85
Fraction V	1.2	3.6	2	61
Initial solution	89.9	70.8	2	100

Separating Components from the Chromium-Aluminum-Iron Tanning Solution. Chemical and Spectral Analysis of Components

To perform chromatographic separation a chromium-aluminum-iron heterocomplex solution of 0.1M concentration was used. Sephadex C25 was used by gradient elution with 1.5M NaCl solution.

Sephadex gel chromatogram of chromium-aluminum-iron tanning solution is shown in Fig. **31**. The five components have the following colors: green, blue, blue-green, magenta and red.

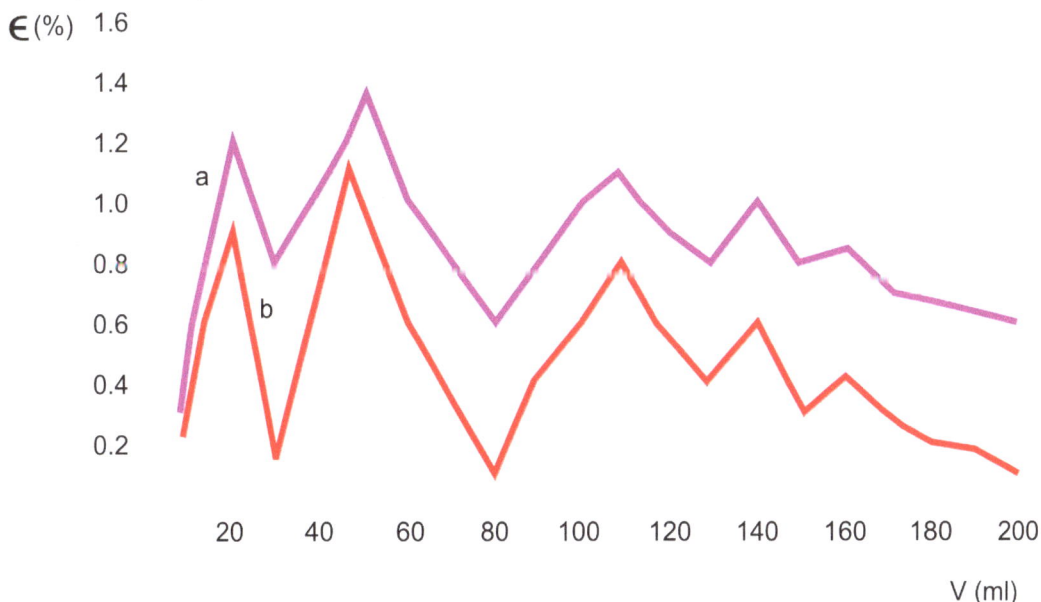

Figure 31: Chromatograms of chromium-aluminum-iron heterocomplex solution (a-λ= 420nm; b-λ= 580nm).

Components were characterized by chemical analysis and spectral analysis. Table **42** presents the results of chemical analyses, and Table **43**, the molar ratios corresponding to the separated components.

Table 42: Chemical analysis of components separated from the chromium-aluminum-iron tanning solution.

Separated component	Cr g/l (moles)	Fe g/l (moles)	Al g/l (moles)	SO_4^{2-} g/l (moles)	$HCOO^-$ g/l (moles)	$C_2O_4^{2-}$ g/l (moles)
I	1.15 (0.021)	-	-	4.10 (0,043)	-	-
II	0.75 0.014)	-	-	0.52 (0.0048)	0,22 (0,0048)	0.48 (0.0051)
III	2.03 (0.038)	-	0.16 (0.006)	1.82 (0.019)	-	2.20 (0.025)
IV	-	1.06 (0,019)	1.02 (0,036)	1.7 (0.018)	-	2.15 (0.036)
V	-	0.95 (0.017)	0.70 (0.025)	1.64 (0.017)	-	2.20 (0.025)

Table 43: Molar ratios - acid groups/M and M/M in components separated from the chromium-aluminum-iron heterocomplex solution.

Component	SO_4/M	HCOO/Cr	C_2O_4/M M=Cr,Fe	Fe/Cr	Al/Cr	Fe/Al
I	2.05	-	-	-	-	-
II	0.34	0,34	0.36	-	-	-
III	0.50	-	0.65	-	0.16	-
IV	1.02 SO_4/Fe	-	1.02 C_2O_4/Fe	-	-	0.5
V	1.00 (SO_4/Fe) 1.47 (SO_4/Al)	-	1.00 (C_2O_4/Al) 1.47 (C_2O_4/Fe)	-	-	1.5

Electron spectra for components I, IV and V are shown in Fig. **32**.

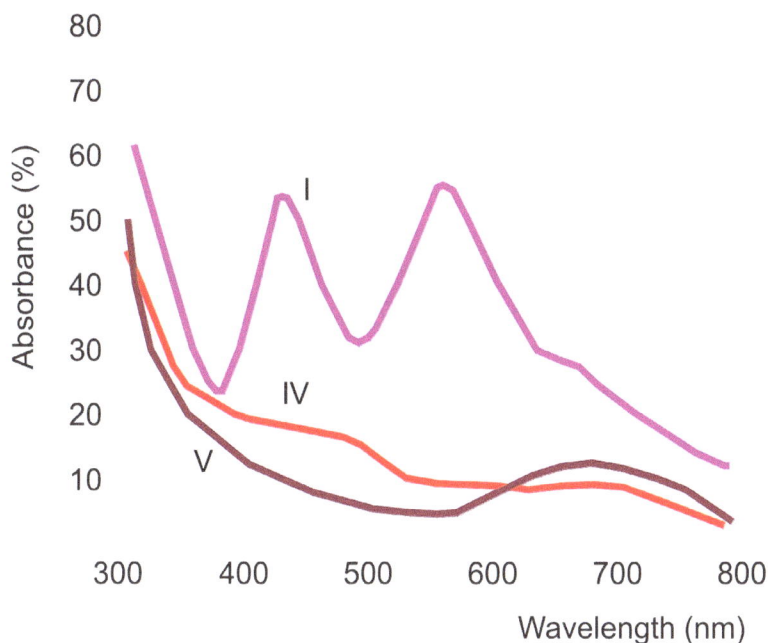

Figure 32: Electron spectra of components I, IV and V from the chromium-aluminum-iron tanning solution.

Electron spectra of components I, II and III in the visible domain are similar and reveal the presence of Cr (III) in weakly distorted octahedral environment.

Electron spectra of components IV and V also include the band at approximately 760 nm, characteristic of Fe (III) in distorted octahedral environment. In the spectrum of component V the band at 760 nm is the only one observed.

Of the components separated from the solution containing chromium, aluminum and iron, only components I and II have been frozen (solid samples were obtained). Other components could not be processed in solid state.

IR spectrum of component I has only bands characteristic to ions SO_4^{2-} and crystallization H_2O, respectively, and IR spectrum of component II has bands characteristic to carboxylate ion ($C_2O_4^{2-}$ and $HCOO^-$, respectively), in addition to bands corresponding to SO_4^{2-} groups.

By Sephadex C25 gel separation five components were found, two components containing only chromium and three components containing Al/Cr and Fe/Al combinations. Separate bicomponents (III, IV, V) have the highest amounts of organic ligands, which confirms the existence of stable anionized components.

Separating Components from the Chromium-Iron-Zirconium Tanning Solution. Chemical and Spectral Analysis of Components

Chromium-iron-zirconium tanning metallic heterocomplex solution was also separated into five parts using Sephadex C25 gel.

Chromatogram of the 0.1M concentration solution obtained by elution with 1.5M NaCl solution is shown in Fig. **33**.

Chemical analysis of component I highlights only the presence of Cr and SO_4^{2-}; component II contains only Cr, and Zr was found only in component IV (with chromium), while in components III and V it appears alongside chromium and iron. Of the five separated components only component I was frozen, the other components remained in the form of glues. IR spectrum of this component closely resembles the spectrum of all I components of the solutions analyzed; they contain bands characteristic to SO_4^{2-} groups and Cr^{3+}.

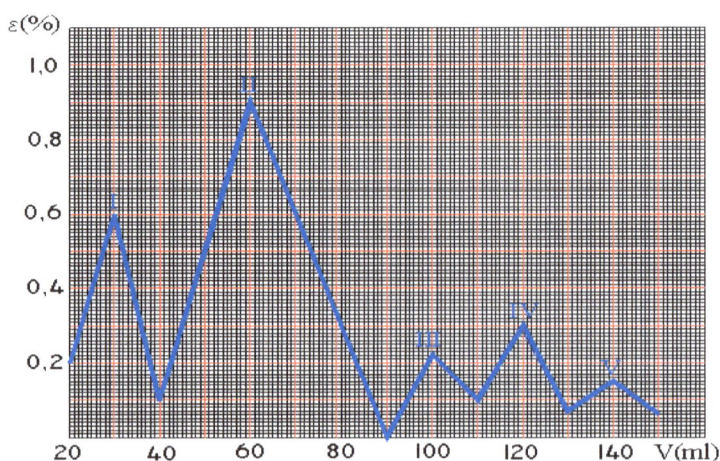

Figure 33: Chromatogram of chromium-iron-zirconium tanning solution on Sephadex C25 eluted with a 1.5M NaCl solution, $\lambda = 420nm$.

Tanning metallic heterocomplexes containing iron are heteropolynuclear complexes; their heteropolynuclearity is achieved through the following ligands: OH^-, SO_4^{2-}, H_2O, $C_2O_4^{2-}$, CH_3COO^- (NIR, IR, UV *etc*).

The presence of Me-O bridges, specific to complex combinations of transition metals, are also present in the case of tanning metallic heterocomplexes. Stereochemistry of the central metals is generally distorted octahedral due to the influence of non-equivalent ligands. Iron presents a distorted octahedral stereochemistry, but a small amount may be located in a tetrahedral environment, so the number of coordination is 4.

In general, the spectral shape of tanning metallic heterocomplexes containing iron is very similar to that of basic chromium complexes. We can, therefore, assume that the type of complexes formed is very little different from the known one. The presence of octahedral structures with configurations such as CrO_6, FeO_6 interconnected through peaks is confirmed by EPR analysis. The stability of these structures is lower compared to those interconnected by edges, but specific to complexes which must have a residual complexing capacity which gives them tanning ability.

The alkali stability scale of heterocomplexes was established as follows:

$$Cr\text{-}Fe > Cr\text{-}Fe\text{-}Zr > Cr\text{-}Al\text{-}Fe.$$

Chromatographic separation analyses have led to the conclusion that tanning heterocomplexes have a stable anionized component, which is extremely important for their use. Its presence correlates well with the fact that large quantities of acids are used in the synthesis, which ultimately anionize the heteropolynuclear structure. Also, the known ability of iron to form complexes with organic residues and identification of a significant quantity of organic residues may explain the stable anionization of part of the heterocomplex solution composition.

Sephadex gel C 25 separations have led to five distinct components, the latter generally consisting of two-component or single-component complexes. Identification, chemical analysis, spectra recorded and the way they interact with protein suggests a selective binding of these components, which, through a cumulative effect, eventually give hydrothermal resistance to collagen. Also, the identification of stable two-component fractions under the specific conditions of the chromatographic technique, is proof of the stability of the complexes and suggests a possible dynamic behavior in interaction with the hide, a wider range of possibilities for interconnection with active groups of the hide.

The varied polydisperse structure and the presence of the stable anionic component may be the causes of favorable behavior in the interaction with hide and may contribute to greater exhaustion of chromium from tanning baths.

Chromatographic separation into heterocomplex systems are original elements of priority and proves the technical, ecological and economical advantages of heterocomplexes use as alternative to chromium tanning [132, 133].

REFERENCES

[1]   Yelmen, H., *2400 Years of Turkish Leather*, Derimod: Istanbul, **2005**.

[2]   Churchill, J.E.,*The Complete Book of Tanning Skins and Furs*, Stackpole Books: Harrisburg,. PA., **1983**.

[3]   Bravo, A. G., Trupke J., *100 000 Jahre Leder*, Birkhäuser Verlag: Basel und Stuttgart, **1970**.

[4]   Deselnicu, M., Olteanu, S., Teodorescu, V., *Istoria prelucrarii pieilor pe teritoriul Romaniei*, Editura Tehnica: Bucuresti, **1984**.

[5]   www.silvateam.com [accessed on 2 October 2012].

[6]   The Iceman's Clothing & Equipment, www.iceman.it [Accessed January 2013].

[7]   Püntener, A, Moss, S., The Iceman and his leather clothes, In: Proceedings of XXVIII IULTCS Congress, Florence, Italy, 2005 March 9-2; AICC Eds, Florence, Italy, [CD-ROM] **2005**, 75.

[8]   Mann, B. R., McMillan, M. M., The Chemistry of the leather industry, **2010**, http://www.scribd.com [accessed on 10 September 2012].

[9]   BASF, Pocket Book for the Leather Technologist, BASF Aktiengesellschaft Ludwigshafen, Germany, **1991**.

[10]   Wolf, G.; Breth, M.; Carle, J.; Igl, G.New developments in wet white tanning technology *J Am Leather Chem As*, **2001**, 96(4), 111-119.

[11]   Siegler, M. Chromium-free tanning process.USA Patent 4,740,211, April 26, 1988.

[12]   UNIDO, Introduction to Treatment of Tannery effluents, Viena, **2011**, http://www.unido.org [accessed on 15th December, 2012].

[13]   UNEP, Tanneries and environment: a technical guide to reduction the environmental impact of tannery operation, Paris, France: UNEP, Industry&Environment Office, **1991**, http://www.unido.org [accessed on 15th December, 2012].

[14]   Pauckner, W. The 1992 John Arthur Wilson memorial lecture: leather tanning in the year 2000, *J Am Leather Chem As*, **1992**, 87(5), 156-174.

[15]   Integrated Pollution Prevention and Control (IPPC) Reference Document on Best Available Techniques for the Tanning of Hides and Skins, European Commission February 2003, http://eippcb.jrc.es [accessed on 10 October 2012].

[16]   Alexander, K. *Developments in clean and eco-labelling, technology in leather industry*. In: Proceeding of Cleantech'95, International Cleaner Technological Seminar; London, UK, 1995; The Leather Technology Centre, London, UK, **1995**, pp.1-7.

[17]   SG –The Test Mark for low pollutant leather products Version 11/2011, pfi-group.org [accessed on September 2012].

[18]   The Toy Safety Council Directive 2009/48/EC, Version 1.0 dd 05/04/2011, http://www.ec.europa.eu/enterprise/sectors/toys [accessed on 15 September 2012].

[19]   Tortora, G.; Calvanese, G. Metalli pesanti nel cuoio: restritioni, metodi di prova ed esperienze analitiche, *Cuoio Pel. Mat. Conc.*, **2008**, 84(1), 3-15.

[20]   Papakonstantinou, D. Research perspectives for the tanning, from beamhouse to tanning, *Cuoio Pel. Mat. Conc.*, **1996**, 72(3), 142-151.

[21]   Badea, N.; Vitan, F.; Maier, S.S. *Chimia si ingineria tabacirii in crom a pieilor*; Casa de Presa si Editura Cronica, Iasi, **1991**.

[22] Abdullah M.A.; Aziz B.K. Equilibrium, Kinetic and Mechanistic Studies of Formation of Cis-mono (AA) bis(oxalato) Chromate 9m) Complex (Where AA is glycine, alanine and histidine) in Moderately Aqueous Acidic Solution, Ibn Al-Haitham *J. For Pure & Appl. Sci.*, **2008**, 21(1), 92-105.

[23] Sanmmarco, U. *Innovating Chrome Tanning Process with Excellent Exhaustion.* In: Proceedings of IULTCS II Eurocongress, Istanbul, Turkey, 2006 May 24-27; Detek Eds, Istanbul, Turkey, **2006**, pp16-17.

[24] Sammarco, U. *Chrome tanning with nearly total chrome fixation.* In: XXII IULTCS Congress, Porto Alegre, Brasil; 16-20 November 1993 ; Revista do Cuoro ABQTIC Proceedings, Porto Alegre, Brasil, **1993**, pp. 468-474.

[25] Marjoniemi, M.; Korpiharju, T.; Mantysalo, E. *In Real Time Monitoring and Control of Chromium Tanning Process.* In: Proceeding of the XIII IULTCS Congress, Friederichshafen, Germany, 1995 May 15-20; Weihert-Druck GmbH, Darmstadt, **1995**, pp.19-28.

[26] Morrera, J.M.; Bacardit, A.; Olle, L. Study of a chrome tanning process without float and with low-salt content as compared to a traditional procedure :Part II., *J.Amer.Leather Chem.Ass*, **2006**,101(12), 254-260.

[27] Covington, A.D. *Chrome management*, Workshop on Pollution Abatement and Waste Management in the Tanning Industry for Countries of the Danube River Basin, Ljubljana, Slovenia, 13-15 June 1995, Technical Report US/RER/95/105; UNIDO: Ljubljana, Slovenia, **1995**.

[28] Kanagaraj, J.; Sadulla, S.; Rao, B.P. High exhaust tanning systems using a novel cross-linking agent (CA), *J.Soc.Leather Technol.Chem.*, **2006**, 90(3), 127-130.

[29] Badea, N.; Maier, S.S.; Badea, C.; Ilincan, D. *The leather tanning with chromium salts complexed with dicarboxilic acids,* In: Proceeding of the X-th Romanian Conference for Textile and Leather, Iassy, Romania, 1992 May 14-16 ; Polytechnic Institute, Textile&Leather Faculty, Iassy, Romania, **1992**, pp.9-15.

[30] Baychrom C-Verfahren, Broscure der Bayer AG, best.Nr.AC.11001, **1992** http://www.lanxess.de [accessed on 5 August 2012].

[31] Knaflic, F. Moglichkeiten zur verringerung der CrIII-Belastung der Nacgerbflotten *Das Leder*, **1990**, 41(4), 61-65.

[32] Costas, D.; Maier, V.; Pruneanu, M.; Sauciuc, S.; Carpov, A. Dicarboxilic Polyelectrolytes with Chrome Ions Bonding Properties used as Ecologic Activators in Pigskin Tanning, *Bulletin of Polytechnic Institute of Iassy*,**1997**,VIII(3-4) 196,196-206.

[33] Holmes, J.M. *Reactive chelators: a new methodology for improved performance of tanning metals.* In: Proceedings of the XIII IULTCS Congress, Friederichshafen, Germany, 1995 May 15-20; Weihert-Druck GmbH, Darmstadt, **1995,** pp.15-21.

[34] Fuchs, K.; Kupfer, R.; Mitchell, J. *Glyoxylic Acid: An Interesting Contribution To Clean Technology.* In: Proceedings of the XIII IULTCS Congress, Friederichshafen, Germany, 1995 May 15-20; Weihert-Druck GmbH, Darmstadt, **1995,** pp14-33.

[35] Fuchs, Kh., Kupfer, R. Glyoxylic acid: an interesting contribution to clean technology, *J.Amer.Leather Chem.Ass*, **1993**, 88(11) p. 402-413.

[36] Moog, G.E.: *Der Gerber. Handbuch für die Lederherstellung.* Ulmer, Stuttgart Germany, **2005.**

[37] Silvestre, F.; Rocrelle, C.; Gaset, A.Reactivity of trivalent chromium salts towards collagen in organic solvent media: nature and role of basifying agents, *J.Amer.Leather Chem.Ass*, **1994**, 88(12) p. 440-451.

[38] Sreeram, K. J. J.; Rao, R., Chandrababu, N. K.; Nair, B. U.; Ramasami, T. High Exhaust Chrome-Aluminium Combination Tanning: Part 1. Optimization of Tanning, *J.Amer.Leather Chem.Ass*, **2006**, 101 (3), 86-95.

[39] Covington, A.D. Theory and mechanism of tanning:present thiking and future implication for industry. *J.Soc.Leather Technol.Chem.*, **2001**, 85(4), 24-34.

[40] Chakraborty, D.; Quadery, A. H.; Azad, M. A. K. Studies on the Tanning with Glutaraldehyde as an Alternative to Traditional Chrome Tanning System for the Production of Chrome Free Leather, Bangladesh J. Sci. Ind. Res., **2008**, 43(4), 553-558.

[41] Covington, D.A.; Chandra, B.; Rao, R.R.; Ramasami, *T. Results TNT-CLRI Programme.* In: Proceedings of the XIII IULTCS Congress, Friederichshafen, Germany, May 15-20 1995; Weihert-Druck GmbH, Darmstadt, **1995**, pp.56-61.

[42] Fennen, J. *Molecular modeling of tanning process.* In: Proceedings of IULTCS Centenary Congress, London, UK, September 11-14 1997; British Section of The Society of Leather Technologists and Chemists, London, **1997**, pp.34-428.

[43] Buttar, D.; Docherty, R.; Tricker, L.; Swart, R.M. *The development of a computer model of collagen and its potential in understanding chrome tannage.* In:Proceedings of IULTCS Centenary Congress, London, UK, 1997 September 11-14; British Section of The Society of Leather Technologists and Chemists, London, **1997**, pp. 43-53.

[44] Reich, G.; Oertel, H. *Fixation of macroscopic collagen structures-a model of tanning reactions.* In: Proceedings of IULTCS Centenary Congress, London, UK, 1997 September 11-14; British Section of The Society of Leather Technologists and Chemists, London, **1997**, pp. 54-55.

[45] Cranston, R.W.; Gleisner, R.W., Macoun, R.G.;Simpson, C.M.; Cowey, S.G.; Money, C.A. *The Total Recycling of Chromium and Salts in Tanning Liquors.* In: Proceedings of IULTCS Centenary Congress, London, UK, 1997 September 11-14; British Section of The Society of Leather Technologists and Chemists, London, **1997**, pp. 225-229.

[46] Faber, K. *Gerbmittel, Gerbung und Nachgerbung*, Umshau Verlag,Frankfurt am Main, Germany, **1984**.

[47] Wehling, B., Luck, W., *Chromium tanning with high exhaustion, Baychrome C*, Bayer AG Broschure, **1987**.

[48] Kochta, J., Slaats; H, Traubel, H.; Wehling, B. Alternatives To Chrome Tannage-The Current Situation, *J.Soc.Leather Technol.Chem*, **1990**, 74(6), 195-210.

[49] Cot, J.; Celma, P.J.; Cario, R.;Cabeza, L.F. *Chrome recovery by ionic exchange resins from the exhausted tanning baths.* In: Proceedings of IULTCS Centenary Congress, London, UK, 1997 September 11-14; British Section of The Society of Leather Technologists and Chemists, London, **1997**, pp. 275-276.

[50] Aloy, M.; Vulliermet, B. *Membrane technologies for the treatment of tannery residual floats.* In: Proceedings of IULTCS Centenary Congress, London, UK, 1997 September 11-14; British Section of The Society of Leather Technologists and Chemists, London, **1997**, pp. 660-666.

[51] Stoica, L.; Constantin, C.; Gaidau, C. ; Lacatusu, I. Removal of Cr(III) ions from tannery aqueous systems, *Journal of Environmental Protection and Ecology*, **2004**, 5(4), p.885-891.

[52] Heidemann, E. *Fundamentals of Leather Manufacturing*, Eduard Roether KG, Darmstadt, **1993.**

[53] Tonigold, L.;Hein, A.; Heidemann, E. Möglichkeiten und Grenzen der Eisengerbung, *Das Leder,* **1990**, V (41), pp.8-12.

[54]   Siegler, M. Contribution toward a tanning procedure-free of chrome *J Am Leather Chem As*, **1987**, 82(3), 117-124.

[55]   Germann, H.-P. High Pressure Injection for Rapid Processing, *Leather*, **199**1,193 (2), 33-34.

[56]   Arbaud, P. The use of aluminium, titanium and magnesium complexes in pretanning, tanning and retanning operations, *J Soc Leath Tech Ch*, **1990**,74 (1), 21-31.

[57]   Celades, R.; Duque, J.; Palma, J.J. Chrome free tanning *J Soc Leath Tech Ch*, **1990**, 74(6), 170-174.

[58]   Gavend, G.; Rabbia, C.; Communal, J.P. The wet-white and dry-option. *Industrie du cuir*, **1991**, 4 (07), 82-93.

[59]   Alain, L.; Hess, M.; Streicher, G.; Puntener, A.; Process for pickling and pretanning raw hides, USA Patent 5360453, January 22, **1993**.

[60]   Sammarco, U. Organic tanning for automobile leathers - limits and perspectives*, Cuoio Pel. Mat. Conc.,* 2005, 81 (4), 247-252.

[61]   Pore, J. ; Jacquinot, E. ; Moretti, J.P. Emploi de nouvelles silices colloidales aux multiples effets dans la preparation des cuirs blancs stabilises *Industrie du Cuir*, **1993**, 7, 67-72.

[62]   Platon, F.; Gaidau, C. *Studies on metallic heterocomplexes synthesis, their characteristics and the way of interaction with dermic substance in view of ecological tanning development* Technical Report, Leather and Footwear Research Institute, Bucharest, Romania, **1994.**

[63]   Gaidau, C., *Research on development of upper leathers from wet-white manufactured leathes*, Technical Report, Leather and Footwear Research Institute (ICPI), Bucharest, Romania, **1995.**

[64]   Platon, F.;Gaidau, C.; Paduraru., G. ; Hanches, R. ; Iacob, E. *Possibilities for chromium pollution elimination from furskin tanning.* In : Proceeding of the Chemistry and Chemical Engineering Conference; 1995 October 20-21; Bucharest, Romania, Politechnica University of Bucharest Eds.: Bucharest, Romania, **1995**, pp. 373-378.

[65]   Gaidau, C.; Platon, F.; Paduraru, G. ; Hanches, R.; Iacob, E. *Chrome free tannage of woolskin.* In: Proceedings of the XIII IULTCS Congress, Friederichshafen, Germany, 1995 May 15-20; Weihert-Druck GmbH, Darmstadt, **1995,** pp.92-95.

[66]   Gaidau, C.; Platon, F.; Paduraru, G.; Hanches, R.; Iacob, E. *Environment protection in furskin tanning industry.* In: Proceeding of the Environment and Industry Symposium, 2003 October 29-31; Bucharest, Romania, Estfalia Eds.: Bucharest, Romania, **2003**, pp.55-61.

[67]   Platon, F.; Gaidau, C.; Sescu, R.Contributions onto tanning with less chrome or free of chrome using some tanning heterocomplexes, *Industria de Pielarie*, **1994**, 4(1), 9-22.

[68]   Gaidau, C.; Platon, F. *Ecological alternative for shoe upper leather tanning-wet-white tanning*; Technical Report 1996/1; Cerpi SA Jubilee Report, Bucharest, Romania, **1996.**

[69]   Gaidau, C.; Platon, F.; Popescu, M. *Consideration on manufacture of we-white leather with mineral tanning materials.* In: Proceedings of IULTCS Centenary Congress, London, UK, 1997 September 11-14; British Section of The Society of Leather Technologists and Chemists, London, **1997**, pp.328-335.

[70]   Gaidau, C; Jitaru, I. *Improvement of chromium tannage by using heterocomplexes with iron, aluminium and zirconium-new aspects and arguments.* In: Proceedings of Light Industry-Chromium in Leather Conference, 2003 November 20-21, Radom, Poland;Politechnica Radomska im Kazimierza Pulaskiego Eds, Radom, Poland, **2003**, pp.65-71.

[71] Gaidau, C.; Platon, F.; Hanches, R.; Iacob, E. *Chrome free tanning of wool skins*. In: Proceedings of Ist International Fur Congress, 1998 May 29 June 1, Kastoria, Greece; GFC-MARC Eds, Kastoria, Greece, **1998**, pp.129-133.

[72] Guthrie-Strachan, J.J. *The investigation, development and characterisation of novel zirconium-based tanning agents*, PhD thesis Rhodes University, October **2005.**

[73] Madhan, B.; Nishad, F. N.; Nair, B.U. Molecular level understanding of tanning using an organo-zirconium complex, *J Am Leather Chem As*, **2003**, 98 (11), 445-450.

[74] Dhayalan, K.; Aravindhan, R.; Sreeram, K.J.; Raghava R.J. Application of rare earth salts for permanent stabilization of skin, *Journal of Scientific&Industrial Research*, **2009**, 68(2), 135-139.

[75] Dasgupta,S. Chrome free tannages: Part I Preliminary Studies, *J.Soc.Leath. Ch.*, **2002**, 86 (5),188-194.

[76] Chandrabose, M.; Fathima, N. N.; Sreeram, K. J.; Rao, J. R.; Nair, B. U.; Ramasami, T. Process for making wet-pink leather. US Patent 20060137101, June 26, 2006.

[77] Anonymus IILF launch for Tanfor T from Kemira, *Leather International*, 2013, 21 January, www.leathermagazine.com [accessed in January 2013]

[78] Platon, C.,F.; Hatmanu, G.; Trisca, R.A.;Daranga, A. ; Moldvai, E. ; Daranga, R. ; Luchian, M. Conclusions on industrial experimentalof chromium-alumnium complexes in leather tanning, *Industria Usoara*, **1989** 10(36), pp.447-452.

[79] Platon, F. ; Gaidau, C. *Development of aluminum-chromium complexes for leather manufacturing (increasing of aluminum proportion in tanning complexes)* Technical Report, Leather and Footwear Research Institute (ICPI), Bucharest, Romania,1989.

[80] Platon, F.; Hatmanu, G.;Trisca-Rusu, A.; Zainescu, G. Studies regarding the chromium-aluminum complexes for leather industry: Jubilee Scientific Session, Iassy, Romania, 10-12 November 1988 ; MEI Sesssion Proceedings 17; IP « Gh. Asachi » Iassy, Romania, **1988**.

[81] Chirita, A. Mineral tanning materials, Polytechnic Institute Publishing House, Iasi, 1988.

[82] Krawiecki, Cz. Conventional method for chrome-aluminium tanning, *Das Leder*, **1980**, LVII (7), 109-113.

[83] Badische Anilin & Soda Fabrik.Complex chromium aluminium acylates. GB Patent 1,265,696, July, 15,1969.

[84] Erdmann H., Miller F.F. Complex basic zirconium salts and aluminum salts. US Patent 4,049,379, September 20, 1974.

[85] Shimenovide, B.S.; Mikhailov, A.N.; Bogdanov, N.V.Formation of complex compounds in solutions with chromium(III) and aluminium, *Kozh-Obuvn. Prom-st.*, **1989**, 40 (5), 3-9.

[86] Gratacos, E.; Marsal, E. Vegetable/Aluminium Combination Tannage, *Leder,* **1990**, 41(12), 242-248.

[87] Chesunov, S.V.; Sankin, L.B.; Makarov, E.; Semyanskiy, Y.Y.Method of producing complex chrome tanning agents, *Tehnol.Legk.Prom-sti.*, **1991**, (34), 101.

[88] Markaryan, S.M.; Tadevosyan, V.A.; Petrosyan, V.M.; Danielyan, V.A. Aluminium sulfate complexes of substituted anilines as tanning agents for skins, *Tehnol.Legk.Prom-sti.,* **1991**, 24, 90.

[89] Bayer AG, Tanning, Dyeing, Finishing, Mineralgerbstoffe, Bayer AG, Leverkusen, Germany,1994.

[90] http://www.performancechemicals.basf.com [accessed on January 2013]

[91] http://www.cromogenia.com/en/productos/76/chrome-tanning [accessed on January 2013]

[92]   Montgomery, K.C. A well defined chromium-zirconium sulphate tanning complex. What future in basic tanning research ?, *J Soc Leath Tech Ch*, **1992**,76, 39-46.

[93]   Karnitscher, T.; Tothne, F.J.; Skaranda, I.T. Investigation of the structure of leather tanned with chrome-zirconium complex tanning material, *J Soc Leath Tech Ch*, **1989**, 68 (10), 8-12.

[94]   Covington, A.D. Leather Tanning Process and Agent. Patent NZ21906,October, 28,1988.

[95]   Quintana, M.; Squitana, G.; Vaira, S.; Arbaud, P.; Reduction or Replacement of Chromium in Tanning, *Bol.Tec.AQEIC*, **1990**, 41(10), 475-485.

[96]   Quintana, M.; Arbaud, P. Reduction or Replacement of Chromium in Tanning, *J Soc Leath Tech Ch*, **1991**, 74(3),101-109.

[97]   Holmes, J., M.; Reactive Chelators: Improving the performance of Tanning Metals, *J Soc Leath Tech Ch*, 1996, 80 (2) 133-135.

[98]   Celades, R.; Duque, J.; Palma, J.J. Titanium tannage, *AQEIC*, **1990**, 41(3), 190-196.

[99]   Pauckner, W. Titan, a tanning agent replacing chrome, *Leather*, **1990**, 4,152-160.

[100] Vychodilova, L.; Ludvik. J. Use of zirconium and titanium salts in leather tanning, *Kozarstvi*, **1991**, 41(7), 190-195.

[101] Castiello, D.; Calvanese, G.; Puccini, M., Salvadori, M.; Seggiani, M.;Vitolo S. *A technical feasibility study on titanium tanning to obtain upper quality versatile leather.* In: XXI IULTCS Congress Valencia, Spain, 27-30 September; ABQEIC Section of The Society of Leather Technologists and Chemists Proceedings, Valencia, Spain, [CD-ROM] **2011**, B44.

[102] Covington, A.D. Leather tanning process using aluminium (III) and titanium (IV) complexes. U.S. Patent 4,731,089, March 15, 1988.

[103] Stephens, L.J. White tannage of woolly sheepskins, *Cuoio*,**1990**, 66(4), 289-304.

[104] Crudu, M.; Ioannidis, I.; Deselnicu, V.; Albu, L.; Crudu, A. *Eco-friendly tanning agents for use in leather manufacture.* In: XXI IULTCS Congress Valencia, Spain, 2011 September 27-30; ABQEIC Section of The Society of Leather Technologists and Chemists Proceedings, Valencia, Spain, [CD-ROM] **2011**, D11.

[105] Peng, B.; Shi, B.; Ding, K.; Fan, H.; Dennis, S. Novel titanium (IV) tanning for leathers with superior hydrothermal stability. III. Study on factors affecting titanium tanning and an eco-friendly titanium tanning method, *J Am Leather Chem As,* **2007**, 102(7),297-305.

[106] Baldi, G.; Niccolai, L. Bilancio economico del sistema di concia con sali di titanio, *Tecnologia Conciarie*, 2002, 153 (2), 84-96.

[107] Anonymous. Concia al titanio, *ARS Tannery*, 2010, 1(2), 48-49.

[108] Covington, A.D.; Lampart, G.S.;Menderes, O.; Farley, R.D.;Murphy, D.M.; Brien, P.O.; Brien, P.O.; Bertini, I.; Fragai, M.; Luchinat, C.; Parigi, G. *Advanced analytical studies of mineral tanning.* In: Proceedings of IULTCS Congress, Cape Town, South Africa, 2001 March 7-10, South Africa Section of SLTC, Cape Town, **2001**, PP051.

[109] Papayannis, A.; Zorn, B. Combinated zirconium, aluminium and chrome tannage. U.S. Patent 3,423,162, Jan 21, 1969.

[110] Kubota, M. Tannages with inorganic compounds-mainly on the iron, and aluminium tannages, *Hikaku Kagku*, **1974**, 20, 1-13.

[111] Ajtulenova, K.T.; Turarva, A.S.; Madiev, U.K.Interaction of iron, zirconium, titanium, and aluminium tanning agents in an aqueous solution, *Kozh-Obuvn. Prom-st.*, **1986**, 105(3), 40-41.

[112] Madiev, U.K. Mineral tanning agents from heteropolynuclear complexes of aluminium, zirconium, and titanium compounds, *Tehnol.Legk.Prom-sti.*, **1984**, 27 (2), 53-57.

[113] Stratev, G.A.; Lebedev, O.P.; Strahov, I.P. Stabilisation of iron tanning material with complexes of some metals, *Tehnol.Legk.Prom-sti.*, **1978**, 4, 55-57.

[114] Keban, M. Bayer AG. Tanning with iron salts - an old system in a new light. Leather International, *Leather Magazine*, **2004**, 3, www.leathermag.com [accessed on 15 September 2012].

[115] Covington, A.D. Tanning Chemistry: The Science of Leather, RSC Publishing, Cambridge, UK, **2009**.

[116] Burger, K. Coordination Chemistry Experimental Methods, AK.Kiado, Budapest, Hungary, **1973**.

[117] Nakamoto, K. Infrared Spectra Of Inorganic and Coordination Compounds, John Wiley Sons., N.Y.,London, **1964.**

[118] Brezeanu, M., Patron, L., Andruch, M. Polynuclear Coordination Compounds and Their Applications, Ed.Academiei Romane, **1986**.

[119] Platon, F., Gaidau, C. *Studies on metallic heterocomplexes synthesis, their characteristics and the way of interaction with dermic substance in view of ecological tanning development*, Technical Report, Leather and Footwear Research Institute (ICPI), Bucharest, Romania,**1991**.

[120] Lagretina, M.F.; Zhadanov, B.V.; Felin, M.G.; Menshikov, B.I.; Strakhov, I.P. IR-spectroscopic study of aqueous chromium (III) tanning solutions stabilized with carboxylic acid salts, *Tehnol.Legk.Prom-sti.*, **1987**, 30(2), 83-86.

[121] Cairus, T. Spectroscopic Problems In Organic Chemistry, vol I, Heydon & Son.Ltd. London,UK, **1967.**

[122] Baker, A.J., Cairus, T., Eghinton, G., Preston, F.J. More Spectroscopic Problems In Organic Chemistry, Heydon & Son Ltd, London,UK, **1967**.

[123] Lever, P.B. Inorganic Electronic Spectroscopy, Elsevier Publishing Company, Amsterdam, London, New York, **1984**.

[124] Cotton, F.A. Metal-Metal Bonding in $[Re_2X_8]^{2-}$ Ions and Other Metal Atom Clusters, *Inorganic Chemistry,* **1965**, 4 (3), 334.

[125] Cotton, F. A. and Wilkinson, G., *Advanced Inorganic Chemistry*, John Wiley and Sons: New York, 1988.

[126] Cotton, F. A.; Walton, R. A. Multiple Bonds Between Metal Atoms Clarendon Press, Oxford (Oxford): 1993.

[127] Montgomery, K.C.; Scroggie, J.G. The isolation of a combined zirconium-chromium(III)basic sulphate complex, **1971**, Aust. J. Chem., 24, 687-696.

[128] Takenouki, K.; Kondo, K.; Nakamura, F. Changes in the chromium complex composition of masked chrome solutions during tannage and affinity of various chromium complexes for collagen, *J.Soc.Leather Techn. Chem.* **1981**, 75,190-196.

[129] http://www.lenntech.com/products/resins/dowex/dowex-ionexchange-resin.htm.

[130] http://www.sigmaaldrich.com/chemistry/chemical-synthesis/learning-center/technical-bulletins/al-142/sephadex-products.html.

[131] Gaidau, C.; Jitaru, I. *Chromatographic separation of some components from complexe tanning solutions.* In: Proceeding of Conferinta de Chimie si Inginerie Chimica, Bucuresti, 1997, October 16-18, Bucuresti, Romania; Bucuresti Politehnica University Eds, 1997, pp.25-31.

[132] Gaidau, C. ; Platon, F. ; Badea, N. Investigation into iron tannage, *J. Soc. Leather Technol.Chem.,* **1998**, 82 (4), 143-146.

[133] European Directive 96/61/IPPC, Reference Document on Best Available Techniques for the Tanning of Hides and Skins, February **2003**, Chapter 6, p.179 (http://eippcb.jrc.es).

INDEX

A

Agents 3, 7-9, 11, 14, 17, 21, 23, 25-9, 32, 36-7, 39, 44-5

Aldehydes 3, 8, 10-3, 21, 27, 97

Aluminum 9-10, 12, 27-9, 33-4, 36-41, 44-6, 50-1, 53-4, 56-61, 64, 70-6, 80, 82-3, 87-91, 93-4, 97, 102-4, 106-9, 117-9

B

Basicity 16-7, 24, 37-9, 41, 50-66, 68-71, 76-9, 85-6, 89-90, 92-3, 110, 116

C

Chromium 3, 6-7, 9-12, 14-64, 66-97, 99-104, 106-122

Collagen 3, 6-9, 14-7, 24, 26-7, 29, 33-4, 36, 44, 49-50, 83, 95, 109-10, 122

Complex 7, 14, 17, 24-6, 31, 33-34, 36-7, 40, 42-3, 46-7, 50, 52, 64-5, 68-9, 81, 87, 91, 96-7, 101-3, 106-7, 116, 121

Crosslinkers 3, 48

E

Ecologic 3, 32, 64

I

Iron 3, 32-4, 36, 41-2, 45, 47-8, 50-1, 55-8, 61-4, 66-95, 97, 99-100, 102-4, 106-9, 111-21

G

Groups 3, 7-8, 14-7, 24-9, 33, 46, 69, 87, 91, 94, 96, 100, 103, 111, 113, 115, 118-20, 122

H

Heterocomplex 33, 36, 38, 40, 43-5, 53-7, 60-1, 63-4, 66-77, 79-80, 83, 85-95, 98-100, 104, 107-11, 113-8, 120-2

L

Leather 3-6, 9, 11-4, 17-28, 31-6, 41, 44-5, 48, 50, 64, 70, 77, 82-3, 94-5, 116
Ligands 15-6, 36, 38-9, 41, 95-6, 99, 102-4, 111, 116, 120-1

M

Mineral 3, 7, 10-2, 14, 20-1, 28, 35, 45, 48

O

Oils 10, 13

P

Process 3-6, 11-2, 15, 17, 19, 22-4, 27-32, 45, 64, 70, 74, 76, 81, 90, 101, 116

R

Reagent 66-7, 73, 80, 88
Retanning 11, 14, 20-1, 25, 27-8, 31-3, 41, 45, 50, 70, 77, 82-3, 94

S

Shrinkage temperature 8-9, 25, 29, 32-4, 40, 48, 117
Synthetic 8, 11-2, 14, 21, 27, 45

T

Tannins 4-5, 8, 11, 14
Tanning 3-17, 19-30, 32-8, 40-2, 44-51, 53, 63-4, 66-7, 70, 74, 76-7, 81-3, 87, 91, 94-5, 99, 102-3, 109-13, 116-22
Titanium 28, 33, 36, 44-6, 51

V

Vegetable 3, 5, 7, 9-12, 14, 21, 32-3, 44-5

Z

Zirconium 10, 27-8, 34, 36, 42-7, 50-1, 58-62, 64, 77-82, 91-4, 97, 101-4, 106-9, 120

www.ingramcontent.com/pod-product-compliance
Lightning Source LLC
Chambersburg PA
CBHW041715210326
41598CB00007B/661